Valuing the Future

Economics for a Sustainable Earth

Valuing the Future:

Economic Theory and Sustainability

Geoffrey Heal

Columbia University Press • New York

Economics for a Sustainable Earth Series
Graciela Chichilnisky and Geoffrey Heal, Editors

Economic forces are driving dramatic changes in the environment of our planet. Our grandchildren may live in a world radically different from our own. Between our lifetimes and theirs, economic activity may cause changes in climate and in plant, animal, and insect populations greater than any since the evolution of human societies, with far-reaching consequences for human well-being. This poses a challenge to economic analysis, for economists have traditionally taken the economy's material and biological surroundings as given, independent of economic activity. Books in this series grapple with the consequences of human domination of global ecological and biogeochemical systems.

Columbia University Press
Publishers Since 1893
New York   Chichester, West Sussex
Copyright © 1998 Columbia University Press
All rights reserved

Library of Congress Cataloging-in-Publication Data

Heal, G. M.
    Valuing the future : economic theory and sustainability / Geoffrey Heal.
        p.   cm. — (Economics for a sustainable earth)
    Includes bibliographical references and index.

    ISBN 978-0-231-11307-6 (paper)
    1. Sustainable development.   I. Series
    HD75.6.H416   1998
    333.7'15—dc21   98–23479

Casebound editions of Columbia University Press books are printed on permanent and durable acid-free paper.

Printed in the United States of America

Contents

Preface   xi

Chapter 1 *What Is Sustainability?*   1

    1.1   History of Sustainability   5
    1.2   Possible Formalizations   7
    1.3   Limitations of Earlier Approaches   11
    1.4   Discounted Utilitarianism   12
    1.5   Sustainability: A Preliminary Definition   13
    1.6   Valuing Environmental Assets   14
    1.7   The Analytical Framework   20
    1.8   Summary   21
    1.9   Outline of the Book   21
    1.10 Conclusions   24
        1.10.1   Conservation   24
        1.10.2   Valuation   25
        1.10.3   National Income   25

PART I SUSTAINABILITY WITHIN A CLASSICAL FRAMEWORK

Chapter 2 *The Classical Formulation*   27

    2.1   Declining Discount Rate   30
    2.2   Conclusion   31
    2.3   Hamiltonians and Adjoint Variables   32

Chapter 3 *Valuing a Depletable Stock*   36

    3.1   Corner Solutions   37
    3.2   Optimal Paths   38
    3.3   The Green Golden Rule   43

3.4	The Rawlsian Optimum	44
3.5	Nonseparable Utility Functions	44
3.6	Summary	45

Chapter 4 *Renewable Resources* — **46**

4.1	Stationary Solutions	48
4.2	Dynamic Behavior	50
4.3	The Green Golden Rule	52
4.4	Ecological Stability	53
4.5	The Rawlsian Solution	55
4.6	Conclusions	55

PART II A BROADER PERSPECTIVE

Chapter 5 *Alternatives to Utilitarianism* — **58**

5.1	Formalizing Utilitarianism	60
5.2	Empirical Evidence	61
5.3	Logarithmic Discounting and the Weber–Fechner Law	62
5.4	Zero Discount Rate	63
5.5	Overtaking	65
5.6	Limiting Payoffs	68
5.7	Chichilnisky's Criterion	69
	5.7.1 Comparison with Overtaking	73
	5.7.2 Constancy of Discount Rates	74
5.8	The Rawlsian Criterion	76
5.9	Discounting Utility or Consumption?	76
5.10	Final Comments	79

Chapter 6 *Depletion Revisited* — **81**

6.1	Optimization with Chichilnisky's Criterion	82
6.2	Conservation and the Chichilnisky Criterion	89
6.3	Differences from Utilitarian Optima	91
6.4	Overtaking	92

Chapter 7 *Renewable Resources Revisited* — **94**

7.1	Constant Discount Rates	94
7.2	Declining Discount Rates	98
	7.2.1 Examples	103
	7.2.2 Empirical Evidence on Declining Discount Rates	103

7.3	Time Consistency		104
	7.3.1	Time Consistency and Chichilnisky's Criterion	107
	7.3.2	Asymptotic Time Consistency	107
	7.3.3	Is Consistency Desirable?	109
7.4	Overtaking		110
7.5	Equal Treatment over Finite Horizons		111
7.6	Summary		112

Chapter 8 *Investment in a Backstop* **115**

8.1	The Model		117
8.2	Utilitarian Investment		118
	8.2.1	Optimality After the Backstop	118
	8.2.2	Optimality When the Stock Is Exhausted	119
	8.2.3	Optimality Before the Stock Is Exhausted	119
8.3	Solving the Problem Recursively		120
	8.3.1	Investing for the Long Run: The Green Golden Rule	122
	8.3.2	The Rawlsian Solution	123
	8.3.3	Investing for Present and Future	123
	8.3.4	Overtaking	125
	8.3.5	Choosing When to Invest	125
8.4	Conclusions		126

PART III CAPITAL ACCUMULATION

Chapter 9 *Exhaustibility and Accumulation* **128**

9.1	The Utilitarian Optimum		129
	9.1.1	Stationary Solutions	131
	9.1.2	Dynamics of the Utilitarian Optimum	132
9.2	The Green Golden Rule		135
9.3	The Chichilnisky Criterion		137
9.4	Conclusion		140

Chapter 10 *Capital and Renewable Resources* **141**

10.1	The Utilitarian Optimum		142
	10.1.1	Stationary Solutions	142
	10.1.2	Dynamics of the Utilitarian Solution	145
10.2	The Green Golden Rule		146
10.3	The Chichilnisky Criterion		150
10.4	Conclusions		152

PART IV POLICY ISSUES

Chapter 11 *Measuring National Income*   **155**

    11.1   Introduction   155
    11.2   Two Concepts of National Income   156
    11.3   A General Model   158
    11.4   Hicksian Income and the Hamiltonian   159
          11.4.1   Constant Discount Rates   159
          11.4.2   Time-Varying Discount Rates   162
    11.5   The Linearized Hamiltonian   163
          11.5.1   Consumption and Hicksian National
                 Income   167
          11.5.2   Hicksian Income and the Discount Rate   168
    11.6   Hicksian Income and Resources   168
          11.6.1   Exhaustible Resources   168
          11.6.2   Renewable Resources   171
          11.6.3   National Income and Capital
                 Accumulation   171
    11.7   Nonutilitarian Objectives   173

Chapter 12 *National Welfare*   **175**

    12.1   A Mathematical Framework   175
    12.2   National Welfare and Resources   180
          12.2.1   National Welfare and Hicksian
                 Income in the Hotelling Case   180
          12.2.2   Exhaustible Resources and National
                 Welfare   182
          12.2.3   Renewable Resources and National
                 Welfare   183
          12.2.4   National Welfare and Capital
                 Accumulation   184
    12.3   Chichilnisky's Criterion   185
    12.4   Sustainable Revenues and National Income   189
    12.5   Summary   193

Chapter 13 *Project Evaluation*   **196**

    13.1   Exhaustible Resources   197
          13.1.1   Initial Shadow Prices   198
          13.1.2   Final Shadow Prices   201
    13.2   Renewable Resources   202

13.3   Sustainable Net Benefits   204
13.4   Investing in a Backstop Technology   205
13.5   Conclusions   206

Appendix   208

A.1   Existence of a Solution for the Exhaustible Resource Case   208
A.2   Existence of a Solution for the Renewable Resource Case   210

References   213

Index   224

Preface

Sustainability is a metaphor for some of the most perplexing and consequential issues facing humanity. These might even include the very survival of our species. Certainly they include the survival, and extension to the rest of the world, of the lifestyle now practiced in the industrial countries. Almost without exception, these issues are rooted in our economic behavior and organization. Yet it is not until very recently that there has been an economic theory of sustainability, or even any systematic application of existing theories to the issues to which it alludes. I see this as a major omission and am puzzled by the slowness of our profession in addressing a set of questions that offer intellectual challenge and policy relevance in abundance. Perhaps they have just not yet been placed clearly on the intellectual map. Whatever the reason, I hope that this book will help to place them more firmly on our research agenda. I hope it will also suggest that economics can contribute to understanding sustainability and that thinking about sustainability can help us to understand economics. Scientific disciplines grow by interacting with problems that stretch them to their limits, and there are many such problems in the complex of environmental problems associated with sustainability.

One of the most intriguing of this set of problems is how we value the future, hence the title of this book. Economists have not really come to grips with valuing events that are centuries away. Typical economic time horizons differ by an order of magnitude from those that are typical for ecological or climatological phenomena. For economists, thirty years is a long time; for scientists concerned with the evolution of the environment, it is short. A lot of what follows is about reconciling these perspectives.

I develop here a framework for thinking about some aspects of sustainability. The framework is one for modeling the dynamic interactions of economic and biological systems, studying the time paths that can emerge from these interactions, and then selecting one or more of these as optimal. I investigate alternative approaches to optimality, inquiring whether there is a concept of optimality that captures the concerns that underlie the emerging interest in sustainability. I suggest that the essence of sustainability lies in three points: a treatment of the present and the future that places a positive value on the very long run, recognition of all the ways in which environmental assets or natural capital contribute to economic well-being, and recognition of the constraints on economic activity implied by the dynamics of environmental assets.

My analysis shows that embodying these concerns in a concept of optimality has important implications for patterns of optimal resource management over time, for the valuation of environmental assets, and for the way in which the use and the services of environmental assets are recorded in national income accounting and indeed for the way concepts such as national income are formalized. Observing these principles leads to more conservative patterns of resource use, higher shadow prices on resources, and a redefinition of several resource-related items in national income accounts.

Much of the work reported here was completed in collaboration with Andrea Beltratti and Graciela Chichilnisky. Geir Asheim and Keisuke Ohsumi have also been unusually generous with their time and helpful with their comments. In addition, I have benefited from conversations with and comments from Yuliy Barishnikov, Michael Hoel, Bill Nordhaus, Charles Perrings, Harl Ryder, Bob Solow, and Jon Strand.

An early version of this manuscript was presented as the Leif Johansen Lectures given at the University of Oslo, March 1995. A later version provided the basis for a series of lectures at the Université Paris X-Nanterre in May of 1997. I am grateful for the invitations to give these lectures, as they provided me with the incentives to complete this project. They also gave me the opportunity to benefit from discussions with the students who attended my lectures. The book was finished during a Fulbright Professorship at the University of Siena.

This is a moderately technical book. Ideally, the reader comes equipped with two types of knowledge. One is knowledge of basic resource allocation theory applied to natural resources, as presented in my earlier book with Partha Dasgupta [41]. Another is a reasonable grasp of the mathematics of dynamic optimization. Despite these ideal prerequisites, the book is close to being self-sufficient in that much of the necessary background is summarized, albeit briefly. The first chapter provides an overview of the issues, methodology, and conclusions, and could serve as a summary for the nontechnical reader. I see the audience as advanced undergraduates, graduate students, researchers and technically oriented policy makers in economics, and graduate students and professionals in other scientific disciplines with an interest in thinking economically about sustainability and environmental conservation in the long run.

Chapter 1
What Is Sustainability?

If a man takes no thought about what is distant,
he will find sorrow near at hand.   (Confucius)[1]

Can existing patterns of human activity safely and sensibly continue unaltered over the long term, or will such continuation lead to unacceptable
consequences? This is the central issue underlying current discussions of
sustainability.[2]

Some of the concerns prompting this question are by now a familiar
part of the daily news agenda. Human consumption of carbon-based
fuels, together with our depletion of carbon-consuming forests, is altering
the natural carbon cycle of the planet, which since time immemorial has
balanced carbon production by animals (humans included) against the
consumption of carbon by plants and microorganisms and sequestration
in the oceans. The disturbance of this cycle is increasing the proportion of
carbon dioxide in the earth's atmosphere, and there is now a consensus
that this is slowly increasing the mean temperature of the planet. We do
not understand fully the consequences of such a change: there seems to be
a chance that for some regions of the world they could be apocalyptic and
irreversible. Such observations lead one naturally to question whether
current patterns of energy use can continue without eventually provoking
unacceptable outcomes: in short, whether they are sustainable.

Similar questions are prompted by the observed loss of biodiversity.
According to distinguished biologists, we are driving species extinct at a

---

[1]Quoted in Newman [84].

[2]This definition, although far from those common in economics, is very close to that
used by Holdren, Daily, and Ehrlich [64], who say "A sustainable process or condition is
one that can be maintained indefinitely without progressive diminution of valued qualities
inside or outside the system in which the process operates or the condition prevails."

rate unparalleled since the demise of the dinosaurs, more than fifty million years ago. These are irreversible, final losses; whatever our technological sophistication, we cannot re-create that which is extinct. The extinction is largely a result of habitat change, and also in some degree a consequence of pollution. Biodiversity is important in many different ways, so again the question arises: are the dimensions of human activity leading to biodiversity loss sustainable? Or will they impoverish us?

A key point is that it is economic forces, economic decisions, that are driving phenomena such as global warming and biodiversity loss. The decision to use fossil rather than solar energy is an economic decision; the decision to use more rather than less energy is also an economic decision. The changes in habitat which lead to extinction are again economically driven; it appears to be more profitable to chop down rainforests and plant coffee or other cash crops than to leave them intact. The choice of polluting rather than nonpolluting technologies is another economic choice. So behind many of the offending dimensions of human activity are economic choices and calculations. We will not significantly change the potentially unsustainable aspects of human activity unless we can develop an economic environment within which they are no longer attractive. In other words, we need to change the rules of the economic game so that it becomes economically rational to pursue sustainable alternatives. A good economic system harnesses private interests in the public good, so that as Adam Smith noted:[3]

> Every individual ... neither intends to promote the public interest, nor knows how much he is promoting it. He intends only his own security, his own gain. And he is in this led by an invisible hand to promote an end which was no part of his intention. By pursuing his own interest he frequently promotes that of society more effectively than when he really intends to promote it.

How could this work? Economic decisions are guided by prices: prices fix the costs of alternative ways of doing business, and the returns from business opportunities. So the phrase the "rules of the economic game" refers to the ways in which prices are determined. We need prices that reflect better the costs associated with nonsustainable policies. This is not a new observation: a long tradition of environmental economics emphasizes the differences between the private and social costs of environmentally harmful activities, and the need to devise economic institutions to close

---

[3][102], book 4, chapter 2, first page.

that gap. In a general sense this book is a contribution to that tradition. We have already made progress in that direction, through institutions such as tradable emission quotas and pollution taxes.

There are, however, several dimensions in which the issue of sustainability is different from, and more demanding than, the earlier issues raised by environmental economics. One is the time dimension. Sustainability is above all about what happens in the long term: about whether we can continue "forever" as we are, and whether the economic rules of the game lead us to make choices that are viable in the long term. Here *the long term* denotes a period much longer than that normally considered in economic analyses, typically at least half a century and sometimes as long as several centuries. These time periods pose a particular challenge for the economists' traditional practice of discounting, and an aim of this book is to consider the alternatives.

A second dimension in which the current concern with sustainability is particularly challenging is that it requires us to address the interactions between our economic systems and a wide range of natural ecosystems. We are coming to realize, in part through the process of losing them, that environmental assets are key determinants of the quality of life in most societies. These assets—forests, clean water, clean air, species, rivers, seas, and many more—are not like physical or financial assets: they are alive and have dynamics, requirements, imperatives of their own. Recognizing this and recognizing that they provide the essential infrastructure for human existence is a key step on the road to building an economic framework that can contribute to the development of sustainable policies. In modeling this framework, one has to draw on the recent literature on ecosystem services and their role in sustaining human societies: the volume edited by Daily [37] is a key contribution here.[4]

My aim here is to review the existing conceptual economic literature on sustainability, and then to develop the concept further within the context of models of the optimal dynamic management of an economy endowed with natural resources. I will use this to suggest that we can give a clear analytical content to the idea of sustainability and can build on this to establish frameworks for project evaluation, shadow pricing, and environmental accounting, all of which are consistent with the underlying theoretical framework, in precisely the same way that current approaches to project evaluation and national income accounting are consistent with and draw their intellectual justification from the

---

[4]See also the book by Baskin [10].

discounted utilitarian approach to optimal growth theory.[5] In the next section I review the existing literature on sustainability, and I also review certain existing concepts that, although not explicitly linked to sustainability, can contribute to the formalization of this concept. Prominent among these are the Fisher-Lindahl-Hicks concept of income as the maximum that we can consume without reducing our wealth and the Meade–Phelps–Robinson concept of the golden rule of economic growth as the configuration of the economy leading to the highest permanently maintainable consumption level.

It is not my intent here to cover all possible interpretations of sustainability or all aspects of a theory of sustainability. My goal is to develop a framework for analyzing sustainability in the context of economic dynamics and of the design and management of economic development strategies. I use a deterministic framework, one that is highly aggregated and simplified, an extension of Solow's classic growth model [103] as modified by Dasgupta and Heal [40]. Though simple, this model has been found by many researchers to yield interesting and robust insights, and the same proves to be true in the present context.

My agenda does not address many aspects of sustainability, some of them unquestionably very important. But one has to start somewhere. Aspects that are central but omitted are those stemming from uncertainty,[6] technical change, and the need to manage the use of global commons or public goods such as the atmosphere and the oceans. Over long time horizons, which are central to discussions of sustainability, uncertainty is pervasive: what will the world look like one century ahead? Two centuries? Technical change is one of the main sources of this uncertainty: in principle, technical change could render many currently threatening practices benign or unnecessary. Economists have often modeled technical change by an assumption of exponentially rising productivity. However, the current problem seems altogether too important to use such a naive approach: any constraint can be avoided in the long run on such a scenario. And although there are models of endogenously generated technical change, we actually know very little about the factors generating enhanced productivity. A satisfactory treatment of these topics will have

---

[5]I am referring to the fact that most of the current practice of cost-benefit analysis has its origins in the works of Dasgupta, Marglin, and Sen [45] and of Little and Mirrlees [77], who took the relatively abstract ideas of the theory of optimal economic growth and applied these to an analysis of the rules governing the use of shadow prices for project evaluation.

[6]See the paper by Asheim and Brekke [9] and the volume by Chichilnisky, Heal, and Vercelli [32], and in particular Beltratti, Chichilnisky, and Heal [14]. See also Chichilnisky and Heal [28] for a nontechnical overview.

to wait, but in the meantime there are aspects of sustainability on which we can make progress.

In the management of the global commons, a key issue is the assignment of property rights in and management of the use of global public goods such as the atmosphere, the oceans, and reserves of biodiversity. Many complex and interesting economic issues arise when one considers how best to manage these. Of course, they are public goods, so we have to be concerned about the possibility of "free riding": they are a very particular type of public goods, namely privately produced public goods. They are privately produced in that the amounts of carbon dioxide or chlorofluorocarbons in the atmosphere are the results of large numbers of decisions made by individuals and firms about lifestyles, technologies, and so on. This introduces an element into the attainment of efficient allocations that is absent from conventional public goods such as defense or law and order, and has interesting implications for the use of tradable permits, a method of establishing property rights and harnessing market forces in the service of the environment that is rapidly gaining attention. In particular, it implies that the initial distribution of property rights among participants in the permit market determines whether the equilibrium attained by the market after trading will be Pareto efficient. These issues are studied in detail in Chichilnisky and Heal [29] and Chichilnisky, Heal, and Starrett [31].

1.1 History of Sustainability

Only recently has *sustainability* become an influential and widely used word. At the 1992 Earth Summit in Rio, considerable attention was devoted to sustainability, and the concept is embodied in the resulting UN Framework Convention on Sustainable Development. In addition, the Organization for Economic Cooperation and Development, the United Nations Committee on Trade and Development, the U.S. Presidential Council on Sustainable Development, and many other domestic and international policy-oriented institutions are devoting time and energy to the analysis of sustainable policies. An environmentalist might find this encouraging. An economic theorist or a public policy economist, on the other hand, could easily find this very worrying, for sustainability is not part of our lexicon; it has no established economic meaning. There is a literature on sustainable development, but this is recent and partial at best, and one certainly could not say that it represents an economic consensus on how to formalize and make operational the ideas associated with sustainability. The concepts and concerns that underlie sustainability are not new. Certainly they go back at least to the 1970s; the Bariloche model (Hererra et al. [62], Chichilnisky [18]) emphasized relevant issues in 1976:

> Underdeveloped countries cannot advance by retracing the steps of...the developed countries.... It would imply repeating those errors that have lead to...deterioration of the environment.... The solution...must be based on the creation of a society intrinsically compatible with its environment. ([62], p. 24)

The concept of "a society intrinsically compatible with its environment" is central: the goal of the literature on sustainability is to understand what this might be and how to implement it. This same model also introduced the concept of "basic needs" as a way of formalizing the minimum requirements needed for successful participation in society and linked the satisfaction of these basic needs with "the creation of a society intrinsically compatible with its environment." Around the same time, the United Nations Conference on the Human Environment in Stockholm (1972) coined the phrase *sustainable development,* and the United Nations Environment Program was founded.

More recently, the Brundtland report [110] produced the following widely-quoted remark: "Sustainable development is development that meets the needs of the present without compromising the ability of future generations to meet their own needs." The timeliness of this report, and the ease with which this phrase rolls off one's tongue, has something to do with the attention given to the concept in recent years. However, this ease is a little misleading: there is no corresponding ease of intellectual assimilation.

Two key concerns are expressed in Bariloche and Brundtland: recognition of the long-run impact of resource and environmental constraints on patterns of development and consumption, and a concern for the well-being of future generations, particularly as this is affected by their access to natural resources and environmental goods. These are an alternative way of articulating the concern that started this chapter, namely whether existing patterns of human activity can safely and sensibly continue unaltered over the long term and whether such continuation will lead to unacceptable consequences.

The framework I develop in the following chapters addresses both of these concerns, which seem very well founded and deserving of explicit recognition and analysis.

1.2 Possible Formalizations

While laying the foundations for modern microeconomics, Hicks [63] defined income as "the maximum amount that could be spent without reducing real consumption in the future." A similar definition can be found earlier in the works of Lindahl [76] and even earlier in those of Fisher [48].[7] Clearly there is a concept of sustainability here: income is defined as sustainable consumption. This has points of contact with the Brundtland report's concern for meeting the needs of the present without compromising the ability of future generations to meet their own needs. It appears that Brundtland may be saying no more than that we, the present, should consume within our income. However, for this to be true, our concept of income would have to be a sophisticated one indeed, encompassing income of all types, psychic as well as monetary, from environmental assets, and adjusting monetary income to allow for the depletion of environmental assets.

This observation naturally raises the question of the appropriate measure of national income and the closely related issue of green accounting. This phrase refers to national income accounting conventions that reflect adequately the services provided by environmental assets and that capture as a cost to society the depletion of natural resources. Developing a satisfactory set of conventions in this area is intimately linked to provision of a satisfactory definition of sustainability: in fact, as we shall see later, the former is the mathematical dual of the latter.[8] A review of the work to date in this field is in Dasgupta, Kriström, and Mäler [44].[9] The interpretations of the Brundtland report's concept of sustainability as consuming within society's income, broadly and greenly defined, should immediately alert us to a possible limitation. Although in general and on average it makes sense to consume within one's income, there are times at which one chooses to consume significantly less in order to consume more at other times. Consuming precisely our income would never allow us to save or dis-save: it would freeze us where we are. Those familiar with the Rawlsian definition of intertemporal justice (see below) will see some semblance here.

---

[7]Nordhaus [87] reviews Fisher's concept of income in the context of sustainability. I occasionally refer to this concept of income as Hicksian, as I believe that this is the way in which it is widely recognized, even though its intellectual origins seem to predate Hicks' use of this framework. See also Nordhaus [86].

[8]*Duality* here is used in the sense of functional analysis: the dual of a space is the set of all real-valued continuous linear functions defined on it.

[9]See also Asheim [5].

The Fisher–Lindahl–Hicks definition of income is often paraphrased as "the maximum consumption that maintains capital intact." In the context of this paraphrase of Hicks, it is natural to mention recent work by Daley [38] and Pearce, Markandya, and Barbier [89], in which they argued for maintaining intact *natural* capital stocks as a condition for sustainability. Sustainable paths for them are paths that maintain intact, in some sense, our stock of environmental assets.

Solow and Hartwick (see Solow [105] and Hartwick [55]) generalized this in the direction of the Fisher-Lindahl-Hicks concept of income and argued that sustainability is captured by a Rawlsian definition of intertemporal welfare:[10] from a Rawlsian perspective, welfare is maximized by maximizing the welfare of the least well-off generation. One can write this formally and succinctly as

$$\max_{\substack{\text{feasible paths}}} \left\{ \min_{\substack{\text{generations t}}} (\text{Welfare}_t) \right\} \qquad (1.1)$$

where Welfare$_t$ denotes the welfare level of generation t, so that we are required by the definition of intertemporal justice in (1.1) to do two things: for any feasible path to find the welfare level of the least well-off generation on that path and then to seek the feasible path amongst all feasible paths that gives the greatest value of this minimal level.

An interesting result by Hartwick and extended by Dixit, Hammond, and Hoel [46], and Solow, shows (unfortunately under fairly strong assumptions) that if a country invests an amount equal in value to the market value of its use of exhaustible resources, then it solves the Rawlsian problem (1.1) and achieves the highest possible level of utility for the least well-off generation. Remarkably, it also achieves the highest feasible constant level of utility given the economy's initial stocks of capital and resources. Investing an amount equal in value to the market value of the use of exhaustible resources is, of course, maintaining intact the value of all capital stocks, including natural capital stocks. In other words, it is living within our Fisher–Lindahl–Hicks income, and a generalization of the Daley–Pearce concept, generalized to allow for the substitution of natural by produced capital of equal market value. As my colleague Graciela Chichilnisky observed,[11] although this result is fascinating and

---

[10]The reference here is to John Rawls' "A Theory of Justice" [92], in which Rawls defined a just society as one so organized as to promote to the greatest extent the well-being of the least well-off group. By analogy, a Rawlsian definition of justice between generations is the program of economic evolution that maximizes the well-being of the least well-off generation.

[11]Private communication.

surprising, it is also slightly suspect from an environmental perspective: imagine all trees replaced by buildings of equivalent value. This maintains the total value of capital stocks intact, yet it is clear that this is not what we mean by sustainable development! Perhaps supply and demand would take care of this problem: as we approach such a situation, the price of trees might rise, and that of dwellings fall, to a point where it is impossible to replace trees by dwellings of equal market value. Such an outcome requires a property such as Walrasian stability of the economy's equilibrium:[12] known sufficient conditions for this are very restrictive.[13] This problem is related to a shortcoming in the existing formalizations of the Rawlsian approach: they value natural resources only as inputs to production, not as assets of value in their own rights. For this reason, these approaches can lead to solutions in which all or most natural capital is replaced by produced capital. I argue below that we have to recognize explicitly all the values of environmental assets, and not just value them as inputs to the productive process.

In the 1960s, Meade [82], Phelps [90], and Robinson [95] introduced the concept of the golden rule of economic growth, which was defined as the configuration of the economy giving "the highest indefinitely maintainable level of consumption per capita." In the standard one-sector neoclassical growth model—the Solow model—this configuration is characterized by equality of the rate of return on capital to the rate of population growth:[14]

---

[12]An economy displays Walrasian stability if the process of prices adjusting proportionally to the difference between demand and supply leads to an equilibrium at which all markets clear. Formally, consider the process $dp/dt = D(p) - S(p)$ where p is a vector of goods prices and $D(p)$ and $S(p)$ are respectively vectors of demands and supplies at the price vector p. Walrasian stability implies that this process has a stable equilibrium at which $D(p) = S(p)$. Few economies seem to satisfy this condition: for more details see Arrow and Hahn [2].

[13]For further development of this point, see Chichilnisky [23].

[14]For those familiar with growth theory, the derivation is simple. Let $C + I = F(K, L)$ where C denotes total consumption, I total investment, assumed to be given by $sF(K, L)$, where K is the total stock of capital and L the labor force. F is the aggregate production function, assumed to show constant returns to scale. Letting $k = K/L$, the rate of population growth be λ and $f(k) = F(K, L)/L$, we have

$$d \ln K/dt - d \ln L dt = dk/dt = sf(k)/k - \lambda.$$

Hence $dk/dt = 0$ iff $sf(k) = \lambda k$. But $C/L = c = (1 - s)f(k) = f(k) - \lambda k$. Hence over paths on which $dk/dt = 0$, c is maximized when $df/dk = \lambda$. The highest indefinitely maintainable level of consumption is attained when the return to investment equals the population growth rate. Note that this level will generally not be immediately attainable from any initial conditions: $df/dk = \lambda$ requires a particular level of capital per head, which may exceed that at the economy's initial conditions.

Rate of return = Population growth rate

The definition of the golden rule as giving the highest indefinitely maintainable level of consumption per capita is clearly another statement about sustainability, but made in a framework devoid of any environmental and resource constraints. The golden rule describes the configuration of the economy giving the highest sustainable utility level.[15] It is natural to extend this concept to dynamic economic models incorporating environmental constraints, and in the following chapters I develop and analyze such an extension, the green golden rule introduced in Beltratti, Chichilnisky, and Heal [11–13].

Note a subtle but very important distinction: finding the configuration of the economy that gives the highest indefinitely maintainable level of consumption per capita is *not* the same as achieving the highest constant level feasible from specific initial stocks of capital and resources. The requirement that a utility level be immediately attainable from the economy's initial conditions is restrictive, and in general rules out the golden rule path or the green golden rule path. This best constant utility path attainable from the initial conditions of the economy,[16] according to Dixit, Hammond, and Hoel, Hartwick, and Solow, is the outcome of investing in produced capital so as to maintain intact the total value of natural and produced capital. This is maximization subject to specific initial conditions: the golden rule is the selection of the configuration that maximizes over all maintainable configurations, independently of initial conditions. Under certain conditions, paths that are optimal in various senses are asymptotic to this configuration.

It emerges from this very brief review that there are clearly elements of formal frameworks that have been available to us for a quarter of a century or more and seem to capture some aspects of what we mean by sustainability. The Fisher-Lindahl-Hicks definition of income and the golden rule are foremost among them. But none of them seem to address fully the concerns articulated in the quotes above from the Bariloche model or the Brundtland report.

---

[15] Actually it describes the configuration giving the highest sustainable consumption level, but as in these models utility depends only on consumption and is increasing in consumption, this is the same as seeking the highest sustainable consumption level.

[16] Under certain specific assumptions about the technology.

1.3 Limitations of Earlier Approaches

Consider each of the approaches just listed. They all have limitations. The Rawlsian approach ties us to the historical accident of initial conditions (see Solow [104] or Dasgupta and Heal [41]): if, as is typically the case in developing countries, the present generation is also the poorest, seeking to maximize the welfare of the least well-off generation does not legitimize a policy of saving now for the future, however great the future returns. The point is that such saving would involve a transfer from the poor present to the presumably richer future, which cannot be sanctioned by a Rawlsian view of justice, oriented as it is solely to the position of the poorest. Yet such saving and capital accumulation is for many the essence of economic development.

As we shall see in detail below, the golden rule does just the opposite of Rawls: it implicitly defines a criterion that disregards the present totally provided we get the long run right, in the sense of attaining the greatest possible maintainable utility level. It could justify a very Stalinist[17] approach to economic development. However, I suggest its use in combination with other elements, in an approach proposed by Chichilnisky [22]. This approach is studied in detail in chapter 5, and its implications are developed at length in subsequent chapters.

Stationarity of natural capital as a criterion, as suggested by Daley and Pearce, has the merit of preserving natural assets. Otherwise it seems arbitrary and places undue emphasis on the status quo, although it does avoid scenarios under which all natural capital is replaced by produced capital of equal value. But given ecological thinking about resilience and spontaneous change in natural systems (see for example Hollings [65]), stationarity seems inappropriate to characterize sustainability. It appears that in the biological world, stillness describes death, not life! Perhaps more important, although the Daley-Pearce approach has the presumably desirable effect of ensuring environmental preservation, it does this by fiat, rather than as a consequence of any deeper general principles. This makes it inflexible, and impossible to use for exploration of the many complex trade-offs that are at the center of economic policy formulation. Indeed, an inability to pose and explore trade-offs is a shortcoming of all of three approaches mentioned here (the Rawlsian, that of Daley and Pearce, and that based on the golden rule). The great merit of discounted utilitarianism, the approach to which we turn next, is that it is flexible

---

[17]The term is Bob Solow's.

enough to allow many trade-offs to be posed and investigated. If, as seems to be the case, we are uncertain of the values that we want to express in our evaluation of environmental projects, this flexibility has great attraction.

1.4 Discounted Utilitarianism

The default criterion for ranking development paths and investment projects, including environmental conservation projects, is provided by the discounted utilitarian framework. Following the approach introduced by Bentham in the nineteenth century, the best path is said to be that which provides the greatest present discounted value of net benefits. Many authors have expressed reservations about the balance that this strikes between present and future. Cline [34] and Broome [17] argued for the use of a zero discount rate in the context of global warming, and Ramsey and Harrod, the founders of the modern theories of dynamic economics, were scathing about the ethical dimensions of discounting in a more general context, commenting respectively that discounting "is ethically indefensible and arises merely from the weakness of the imagination" and that it is a "polite expression for rapacity and the conquest of reason by passion" (see Ramsey [91], p. 543; Harrod [53], p. 40; and Heal [60]).[18] It is ironic that a practice so roundly condemned by the founders of intertemporal economics has come to occupy so central a position in the field. It may be fair to say that until now discounted utilitarianism has dominated our approach more for lack of convincing alternatives than because of the conviction it inspires. It has proven particularly controversial with noneconomists concerned with environmental valuations.

The legitimacy of discounting is a complex issue, and the comments of Ramsey and Harrod, though perceptive and pointed, do not do it justice. As we shall see in chapter 5 (see also Heal [58]), discounting of future utilities is in some sense logically necessary; without it one encounters a variety of unsettling paradoxes. The distinction between discounting future utilities in the evaluation of development programs and the discounting of future benefits in cost-benefit studies has also to be borne in mind (Heal [60]). Discounting future utilities does not necessarily imply that it is appropriate to discount future benefits in cost-benefit studies.

---

[18]Heal [60] argued that a zero consumption discount rate can be consistent with a positive utility discount rate in the context of environmental projects.

A positive utility discount rate forces a fundamental asymmetry between the treatments of, and the implicit valuations of, present and future generations, particularly those very far into the future. This asymmetry is troubling when dealing with environmental matters such as climate change, species extinction, and disposal of nuclear waste, as many of the consequence of these may be felt only in the very long run, a hundred or more years into the future. At any positive discount rate these consequences will clearly not loom large (or even loom at all) in project evaluations. To illustrate, if one discounts present world GNP over two hundred years at 5% per annum, it is worth only a few hundred thousand dollars, the price of a good apartment. Discounted at 10%, it is equivalent to a used car.[19] On the basis of such valuations, it is irrational to be concerned about global warming, nuclear waste, species extinction, and other long-run phenomena: the long-run future is irrelevant! Yet societies obviously *are* worried about these issues, and are actively considering devoting very substantial resources to them, as noted in the opening section of this chapter. So a very real part of our concern about the future is clearly not captured by discounted utilitarianism. Because economic institutions and procedures should capture society's values and concerns, we have to find a more satisfactory alternative. There is interesting empirical evidence (see Lowenstein and Thaler [80]), to which we return in chapter 5, that individuals making their own decisions do not compare present and future by discounting the future relative to the present at a constant discount rate, as is standard in the discounted utilitarian approach. Rather, they seem to apply a discount rate that varies with the time horizon, and is quite high over short periods (15-20% over a few years), but falls rapidly with the length of the horizon under consideration, being as low as 2% for horizons of several decades.

We need a framework for considering very long time horizons that is sensitive to both present and future, and possibly more consistent with the approach implicit in individual choices. This is driving an interest in formalizing the concept of sustainability, and the associated unease with the hitherto standard economic framework based on discounted utilitarianism.

1.5 Sustainability: A Preliminary Definition

We have now outlined earlier approaches to sustainability, their limitations, and the intuitions and concerns behind this concept.

---

[19]These are at New York prices.

The time has come to build on this. I suggest here, and argue in detail below, that the essence of sustainability lies in three axioms:

- A treatment of the present and the future that places a positive value on the very long run
- Recognition of all the ways in which environmental assets contribute to economic well-being
- Recognition of the constraints implied by the dynamics of environmental assets.

The first of these points is captured in a definition of sustainability proposed by Chichilnisky [22], and by the green golden rule and the overtaking criterion with a zero discount rate. The second point relates to the way in which we value environmental assets. It implies that we recognize all the dimensions of their value.

1.6 Valuing Environmental Assets

Finding a framework that allows for a complete integration of the ways in which environmental assets contribute to the economy is a central part of the research agenda. Unfortunately systematic research has only begun to scratch the surface of this complex and far-reaching issue. However, it has already shown that there are myriad ways in which environmental assets are sources of value.[20]

Environmental assets are valuable as sources of knowledge; this is one of the sources of value in biodiversity, the source that is tapped in biological prospecting and in the famous Merck-InBio deal.[21] The point here is that many pharmaceutically valuable products and many agriculturally valuable crop strains have been developed from species found in the wild. This point cannot be overemphasized: according to a report to the National Academy of Sciences of the United States, one third by value of the pharmaceuticals sold in the United States, over $60 billion in current market value, were originally obtained from plants or insects, and many of the more robust grain species have likewise been derived from specimens found in the wild. Obtaining pharmaceutically valuable products from plants, insects, or animals, largely from those in rainforests, is now recognized by most major drug companies as a

---

[20]The most comprehensive compilation of the values of environmental assets is in Daily [37]. For a review of some of the underlying biology, see Baskin [10].

[21]For details of this and similar deals, see Chichilnisky [21] and Chichilnisky and Heal [30].

commercially valuable strategy, as is the systematic study of traditional healing methods, which sometimes give clues to pharmacologically active compounds. A significant number of pharmaceutical companies now actively investigate the medical efficacy of traditional remedies, almost all of which are based on extracts from plants and insects, and some major academic medical centers—the College of Physicians and Surgeons at Columbia University to name one—have set up departments of traditional medicine. It is hard to place even a rough value on biodiversity as a source of medical and agricultural information, but this is clearly a very large value indeed. In this sense biodiversity is a capital stock yielding a flow of services of at least several tens of billions of dollars annually. Little is new in this insight, incidentally: the recognition of the pharmaceutical value of biodiversity is simply a recognition of value in traditional healing practices. Societies have traditionally used plant and insect derivatives to treat illness; Shakespeare recognized this in *Romeo and Juliet:*[22]

> O! mickle is the powerful grace that lies
> In herbs, plants, stones and their true qualities:
> For nought so vile that on earth doth live
> But to the earth some special good doth give,
> Within the infant rind of this weak flower
> Poison hath its residence and medicine power.

Biodiversity also plays a critical insurance role. Its contribution to the maintenance of rice production provides an excellent example. Rice is one of the most important food crops in the world, and certainly the most important in Asia. Strains of wild rice are preserved as a source of genetic variation by the International Rice Research Institute (IRRI) in the Philippines, and the returns to this investment have been almost incalculable. In the early 1970s a virus called the grassy stunt virus posed a major threat to Asia's rice crop: it was expected to destroy 30-40% of the crop, bringing great hardship and economic loss. The threat was avoided by genetic engineering in which an immunity-conveying gene from wild rice was transferred to commercial varieties. For the record, the strain of wild rice containing the ability to resist grassy stunt was located in the wild in only one place, a valley that was flooded by a hydro power dam shortly after it was collected. If the IRRI had been set up a

---

[22]*Romeo and Juliet,* II:iii. Cited in Gurr and Peach [51], which is an interesting and informative review of developments in plant-based pharmacology.

few years later, we would have lost the ability to resist that virus and consequently lost much of the Asian rice crop.

A similar event occurred in 1976: another threatening disease was defeated by genetic manipulation that transferred to commercial varieties the immunity carried by certain strains of wild rice. So the investment in conserving genetic diversity has been responsible for maintaining the productivity of Asian rice farming.[23]

Environmental assets also have value as life-support systems: green plants produce the oxygen without which animals die. In fact, the planet's atmosphere originally contained no oxygen: oxygen was introduced by plants and microorganisms. Bacteria not only clean water but create and fertilize soil.[24] Insects pollinate plants and control agricultural pests.[25] All these activities are absolutely crucial in the maintenance of human life. Furthermore, the extent of these life-support activities is certainly not fully understood; recent theories hold that the destruction of ecosystems is contributing to the spread of new diseases to humans and to increases in the geographic range of traditional diseases.

On less speculative matters, it has been discovered that damage done to soil ecosystems by pollution has reduced their ability to purify water collected in aquifers and reservoirs, and so increased the need for expensive water purification plants. In cases such as this, it is well within the limits of our present methodology to assign a value to the services provided by environmental assets. For example, we can easily asses the cost of destroying the microorganisms that purify water as it passes through the soil. It is the cost of building and running water purification plants to provide the services once provided by the ecosystems. For the watershed of a large urban area, this can require an investment of tens of billions of dollars. This investment must be repeated as equipment wears out every few decades, with running costs in the interim. And even after all this expenditure, chemical purification is less satisfactory than the natural alternative. To give a concrete example, New York's water comes from a watershed in the Catskill mountains. Until recently the natural purification process, carried out by microorganisms in the soil, was sufficient to cleanse the water to Environmental Protection Agency standards. In recent years, sewage, fertilizer, and pesticides in the soil reduced the efficacy of this process. As a result, the city was faced with a choice: it could either restore the integrity of the Catskill ecosystems that purify

---

[23]This example is taken from chapter 14 by Myers in Daily [37].
[24]See Daily, Matson, and Vitousek, chapter 7 in Daily [37].
[25]See Naylor, Ehrlich, and Ehrlich, chapter 9 in Daily [37].

water or build a purification plant at a capital cost of $6–8 billion (about $1000 per adult in New York), plus running costs of the order of $300 million annually.

Restoring the integrity of the Catskill watershed meant buying land in and around the watershed so that its use could be restricted, and subsidizing the construction of better sewage treatment plants. The total cost of measures needed to restore the functioning of the Catskill watershed came to $1–1.5 billion. The cost of attaining the same outcome through a purification plant was $6–8 billion, an immense difference. So in this case investing $1–1.5 billion in the environment saved an investment of $6–8 billion in physical capital.[26]

In addition to these immensely practical values, environmental assets such as animals, plants, and even landscapes may have an intrinsic value, a value independent of their anthropocentric value, and they may have a right to exist independently of their value to humanity.[27] It is hard, if not impossible as a matter of principle, to place an economic value on such values and rights. Perhaps respecting them has to be seen as a constraint on society's economic activities, and we should not seek to trade them off against other goals.

Biodiversity may have great cultural significance for certain societies. This role is not restricted to traditional societies; examples are the bald eagle as a symbol of the United States, the bear as a symbol of Russia, or the relationship between Hinduism and the elephant. Biodiversity plays a crucial role in many social and cultural traditions and in forms of artistic expression. Many dimensions of expression, and many elements of social and cultural self-awareness, could be damaged by extensive loss of biodiversity.

All of these sources of value are critically important, as are others that have not been mentioned. However, there are distinctions. Information value and life support value are instrumental values and may vanish if we find synthetic substitutes for natural resources in these roles. They value the environment as a means rather than as an end. This is in contrast to recognizing a value intrinsic in environmental assets, irrespective of instrumental values.

Explicit recognition of all the contributions made to human societies by environmental assets seems an essential element of the concept of

---

[26]For more details on this example, see Chichilnisky and Heal [30].

[27]A discussion of some of these philosophical issues can be found in Kneese and Schultze [69], Murphy [83], and Rolston [96]. A fascinating recent paper by Ng [85] is also recommended.

sustainability. How do we represent this analytically? In the economics literature on the optimal use of natural resources, the standard approach has been to include in the model a stock of a natural resource whose value at time t is denoted s_t, and to indicate that it is the consumption, the depletion, of this stock that contributes to welfare. Letting c_t be this consumption rate at date t, utility, production, or both are shown to depend on c_t. The dynamics of the natural resource also reflects its rate of consumption, so that for an exhaustible resource

$$\frac{ds_t}{dt} = -c_t$$

and for one that is renewable and whose growth is a function $r(s_t)$ of the current stock one has

$$\frac{ds_t}{dt} = r(s_t) - c_t$$

It is a clear implication of the literature on ecosystem services (see Daily [37]) that this formulation is inadequate, and that in many important cases the impact of environmental assets on economic well-being is a function at least as much of the existing stock as it is of the rate of consumption of that stock. Examples come to mind readily. A forest contributes to economic well-being by providing timber; the level of this contribution is measured by the rate of consumption of the forest. It also contributes to well-being by removing carbon dioxide from the atmosphere and producing oxygen. Here the contribution is a function of the size of the stock. A forest also contributes to well-being by acting as a climate stabilizer, as a shelter and support for biodiversity, and in many cases as a watershed, managing water flows and purifying water. In all of these roles, the extent of the benefit provided is a function of the stock. Of course, stock is a simplification here: the distribution of trees by age and species certainly matters, so that for a detailed representation stock should be a vector reflecting these characteristics.

Biodiversity is an environmental asset that contributes to economic welfare entirely as a function of its stock, rather than as a function of its rate of consumption. Viewed economically, biodiversity is an exhaustible resource: its depletion via species extinction or other loss of genetic diversity is irreversible, at least on human time scales. Thinking of depletion of biodiversity as consumption, we can see that there is no direct benefit from such consumption, although it may be a by-product of other valued activities. The benefits that we gain from biodiversity are a function of the extent of this diversity, suitably measured (i.e., of the size of the stock).

In the case of soil, what matters from the perspective of economic welfare is probably a combination of physical stock and productive quality. The physical stock certainly can be depleted via erosion by wind or water, and the productivity of a stock can be reduced by overcropping or misuse of agricultural chemicals.

Other cases are probably much more complex. In some cases, what gives the maximum contribution to economic well-being is a fully functioning, noncompromised and nondegraded ecosystem. The extent to which there are simple indicators of this state, whose dynamics can be related to economic activity, is an important and largely open research topic.

In my analytical modeling, I summarize this complex discussion quite crudely: in general I assume that utility is derived from a flow of consumption that can be produced from the environment and from the remaining stock, so that society's instantaneous utility at each point in time can be expressed as a function $u(c_t, s_t)$, where c_t is a flow of consumption at time t and s_t an environmental stock at that date, as in Krautkraemer [71] and subsequently in Beltratti, Chichilnisky, and Heal [11]. Of course, this is a heroic oversimplification. However, it is in the best traditions of growth theory, indeed economic theory in general, to take the first steps in exploring new issues by treating them in a very aggregative manner.

A natural next step in environmental research is to study more closely the technologies and processes by which the stock of an environmental asset provides value to the community. For example, in the case of biodiversity, the services are not a function only of the stock. If one thinks of the service of provision of knowledge, as discussed above, then the extent to which this service is provided depends on the stock and also on the resources allocated to biological prospecting. If one thinks of the recreational services of a forest, which clearly depend on the forest stock, then these also depend on transportation possibilities and accommodation possibilities. In general, for each category of service provided by environmental assets, there are processes by which this service is provided and complementary inputs involved in the provision of the service. Aggregative models of the type I review here do not and are not designed to capture these details; they are intended to provide insights at a more general level about the use and conservation of environmental assets.

1.7   The Analytical Framework

The remainder of this book proposes an analytical framework that is conceptually quite simple. The framework is one for modeling the dynamic interactions of economic and biological systems, studying the time paths that can emerge from these interactions, and then selecting one or more of these as in some sense optimal. Optimality is defined with respect to the current and long-term behavior of the path, and of course placing value on the contributions of environmental assets to economic well-being.

Symbolically, let e be a vector of economic variables and b be a vector of biological or ecological variables. Each has a dynamic that depends on its own values and those of the other:

$$\frac{de}{dt} \equiv \dot{e} = f(e,b), \quad \frac{db}{dt} \equiv \dot{b} = g(e,b) \qquad (1.2)$$

Of the paths satisfying the equations (1.2), some are sustainable and some not. A precise definition of sustainability will come later. Typically a necessary condition for a path to be sustainable is that the path is contained in a region that is bounded away from some of the coordinate axes: it is bounded away from some axes because we do not want some key variables to go to zero. These would certainly include the stocks of various biological assets. What kinds of paths stay in a region that is bounded and is also bounded away from some of the axes? Clearly nonzero stationary solutions to (1.2) meet this condition. Limit cycles can also meet this condition. So can chaotic attractors. So mathematically we are looking for stationary, periodic, or semiperiodic solutions to (1.2); these are the paths that might be sustainable, in the sense that they can be continued for ever without certain key variables going to zero. This framework captures the constraints imposed on economic activity by the environmental asset base. Within the universe of such paths, we then want to find those that are in some sense best. We need to devise a way of ranking the alternatives and then optimize according to this ranking. If we were to focus only on stationary solutions as possible sustainable paths, this would be a finite dimensional optimization problem: the analysis associated with what I call the green golden rule below falls into this category. More generally, it is a more complex infinite dimensional optimization problem.

1.8 Summary

If we value the long run adequately, recognize the values of environmental assets, and recognize the constraints imposed on economic activity by their dynamics, then everything else that is generally associated with the concept of sustainability is logically implied. This approach will lead to the selection of what we think of intuitively as sustainable policy options when considering policies toward such complex long-term issues as global warming, species preservation, and the management of nuclear waste.

The key link in this argument is that valuing the longrun implies substantial concern for social costs that will occur one hundred years or more ahead, and hence a concern about the long-run consequences of climate change, nuclear waste disposal, and loss of biodiversity. Such concerns are rational only if we place more weight on the very long run than is consistent with the usual approach, namely discounted utilitarianism. So although valuing the longrun, valuing environmental assets, and respecting the constraints they place on us do not themselves describe precise sustainable policies on issues such as climate change and biodiversity, they are necessary conditions for the systematic and consistent selection of sustainable policies as optimal.

This framework is consistent with the attitudes of Bariloche, Brundtland, Daley, Pearce, Solow, and the other authors mentioned earlier, namely that sustainability is about intergenerational equity, resource constraints, and concern for the impact of human activity on the environment over the long run.

1.9 Outline of the Book

In the following chapters, I introduce and develop a framework for discussing the concept of sustainable development. I start from the premises already introduced, that there several simple but central elements to sustainability. One is placing sufficient value on the very long run, which means placing on the very long run a greater value than is done by the discounted utilitarian framework. Valuing the very long run appropriately is essential if we are to analyze issues such as global warming and the disposal of nuclear wastes, both of which may pose threats to human societies over very long time horizons, horizons running into hundreds of years. A second key element is recognizing explicitly the several sources of value associated with stocks of environmental assets, and in particular recognizing that these stocks may be valuable in their own right, and not just for the consumption goods they can yield. A third key element is recognizing the constraints placed on economic possibilities by the growth patterns of environmental resources.

I introduce simple dynamic economic models within which one can examine some of these matters. One model allows us to study the optimal use patterns for a resource that is exhaustible, but whose stock is a source of value. The second allows us to conduct the same exercises for a renewable resource; in each case we can enquire into the optimal use of the resource over time, given that its stock is a source of utility. Finally, we look at the problem of making an investment whose payoff may be very, very far into the future. Examples are investing to ensure the safe disposal of nuclear waste, investing to prevent global warming, or investing in a technology to produce a renewable alternative to fossil fuels.

Within each of these three scenarios, I study the optimal policy, the associated social valuations or shadow prices, and their implications for national income accounting and project evaluation. I do this using several alternative definitions of optimality: the conventional discounted utilitarian definition, a concept of optimality that involves attaining the maximum sustainable utility level and is a generalization of the golden rule of neoclassical growth theory (the green golden rule), a concept of optimality due to Chichilnisky that treats the present and the very long-term future quite symmetrically, and, when appropriate, the concepts based on the idea overtaking and on the intertemporal version of Rawls's concept of justice. The discounted utilitarian approach is present oriented, maximizing sustainable utility leads us to focus entirely on the very long run, and Chichilnisky's criterion combines aspects of both. Each of these three concepts of optimality is explored in each of the three modeling scenarios. Certain general results emerge from this exploration, particularly with respect to the impact of valuing stocks and valuing the longrun on shadow prices, project evaluation, and the appropriate definitions of national income and net national product.

I have arranged the material in the book in four parts: "Sustainability Within a Classical Framework," "A Broader Perspective," "Capital Accumulation," and "Policy Issues." The first part essentially applies the utilitarian approach to the set of models outlined above; this is supplemented by analyses of the implications of the golden rule and Rawlsian frameworks. It shows that even within this framework, there are interesting insights into sustainable management of natural resources. Recognizing that natural capital contributes to human welfare and that it has its own dynamic takes us a long way in the direction of selecting policies that accord with our intuition about sustainability.

In "A Broader Perspective" I try to come to grips with the insistence of Ramsey, Harrod, and their predecessor, Sidgwick, that future utilities should not be discounted. We look at Ramsey's stratagem for avoiding discounting, von Weizäcker's concept of "overtaking" as a generalization of Ramsey's program, and Chichilnisky's criterion, the only one to provide a framework as watertight and comprehensive logically as the utilitarian while going some way toward the Ramsey-Harrod-Sidgwick goal of treating present and future equally. Pursuing the implications of this criterion leads us into difficult problems technically and conceptually: technically because standard optimization techniques are not applicable, and conceptually because we are forced to grapple with some complex issues of dynamic welfare economics. However, these difficulties are rewarding: we narrow down more precisely the most productive formalizations of the concept of sustainability.

The chapters in "Capital Accumulation" show that none of the key conclusions of the previous parts depend on a key simplifying assumption of the model, the absence of produced capital. They revisit the earlier analysis in an economy in which produced goods can be invested in forming capital goods that to some degree may substitute for natural resources.

"Policy Issues" makes a preliminary attempt to set out the implications of all of the previous work for two key issues: how to measure national income in a dynamic resource-using economy (the issue often called green accounting) and how to pick the shadow prices used to value environmental assets in such an economy. In the discussion of national income accounting, I note that the approach most widely pursued to date, namely that based on the work of Fisher, Lindahl, Hicks [63], and Weitzman [109], has a quite different conceptual basis from that underlying the static welfare economics definition of national income. As noted above, a concept of sustainability is built into the Fisher-Lindahl-Hicks definition of income, a concept applied by Hicks to individual rather than national income. The more traditional welfare economics definition of national income, when extended to a dynamic framework, is quite distinct from this approach and is closer to a measure of national wealth than to one of national income.

1.10 Conclusions

Conclusions are usually found at the end of a book. I am placing them here because I want the nontechnical reader who dips into this book and does not go far beyond this introduction to be able to take away a sense of the destination we reach as well as the route we follow.

Many intellectually rigorous approaches to environmental management imply policies that are very conservative of natural capital. They range from approaches that are minor modifications of the standard discounted utilitarian approach to others that are radically more future oriented, such as the green golden rule. All produce optimal paths that could be implemented by decentralized behavior in a market economy, given suitable prices. And all have associated with them measures of national income that embody and reflect their underlying philosophies. The choice between them is ultimately a matter of social and individual values; whatever these are, within the range considered here, there is an economic framework to implement them.

These consequences of the alternative approaches can be listed under three categories: the effect on conservation of natural capital, the valuation of natural resources, and the measurement of national income.

1.10.1 Conservation — The smallest change we consider in the standard discounted utilitarian approach is including the stock of natural capital in the utility function, so recognizing that the stock of natural capital can affect human well-being directly. This is a part of the message from the ecosystem services literature. This move alone changes qualitatively the nature of an optimal path. For nonrenewable resources, which the standard approach would deplete completely, this alternative leads to situations where a portion of the initial stock of natural capital should be conserved indefinitely (chapter 3). The size of this portion depends on the discount rate. For renewable resources, it leads us to conserve a stock that exceeds that giving the maximum sustainable yield, improving the resilience of the system (chapter 4).

Going further and changing the optimality criterion in the direction of the green golden rule or Chichilnisky's criterion leads to still more conservative recommendations. The green golden rule can justify conservation of the entire initial stock of a nonrenewable resource, as will a Rawlsian approach. Chichilnisky's criterion can justify a significantly more conservative approach than usual, without going as far as the green golden rule (chapters 6 and 7).

1.10.2 Valuation — Again, simply recognizing that natural capital stocks can affect welfare directly has a significant effect: it leads to higher shadow prices or social valuations of natural resources. The other alternatives considered reinforce this effect. More future-oriented criteria lead to higher social valuations.

In some cases, the time paths of resource use that are optimal will not maximize the present value of profits at the associated shadow prices. Instead, they maximize sustainable profits, or a combinations of present value and sustainable profits (chapters 12 and 13). Maximizing sustainable profits means attaining the highest level of profits that can be maintained indefinitely: it is the equivalent for profits of the green golden rule.

1.10.3 National Income — Recognizing the importance of natural capital and of using more future-oriented criteria than the present value of utilities leads to significant departures from the standard measures of national income (chapters 11 and 12). In the context of environmental issues, researchers have pursued two different approaches to defining and measuring national income. One is based on the insights of Fisher, Lindahl, and Hicks, and has an intuitive consonance with sustainability. Another is based on the national income concept used in welfare economics and resource allocation theory. The two are distinct and involve different expressions for national income. The latter is more general and has variants that can be associated with each of the different approaches we consider, including in particular those other than discounted utilitarianism. It measures welfare in the same spirit as the standard national income measure of welfare economics, and in a dynamic context corresponds more to a concept of wealth than to one of income.

Part I
Sustainability within a Classical Framework

Chapter 2
The Classical Formulation

We consider first the simplest and most classical formulation of the problem of the optimal management of a natural resource. This formulation assumes the resource to be exhaustible and simplifies the analytical framework by neglecting two of the issues that will be important in coming to grips with sustainability: placing adequate value on the long-run future and recognizing all sources of value from environmental assets. However, it does recognize, albeit in a rather basic way, the constraints imposed by an exhaustible resource on consumption patterns. It provides analytical building blocks that become foundations for some of the subsequent structures; it also helps to develop appropriate intuitions.

One could think of this as a model of the depletion of an oil or gas reserve or reserves of other minerals. These are finite in amount, nonrenewable, and of no value other than as inputs to production. Despite its limitations, this formulation, introduced by Hotelling [66] in 1931, is instructive. It shows clearly the forces at work in the discounted utilitarian approach, provides a simple framework for the presentation of mathematical techniques that are central to the later work, and is a building block in the construction of more general and more satisfactory frameworks. The story it tells recurs as a subplot in the more complex dramas of models that capture more of our real concerns.

The problem in this framework is that of choosing the time pattern of use of the exhaustible resource so as to maximize the integral of the discounted utilities obtained from consumption of the resource, subject of course to a constraint that the total amount of the resource used over time should not exceed the initial stock; symbolically, the problem is

$$\max \int_0^\infty u(c_t)e^{-\delta t}\,dt \quad \text{s.t.} \quad \int_0^\infty c_t\,dt \le s_0 \qquad (2.1)$$

Here $u(c_t)$ is a utility function that is assumed throughout this book to be increasing, strictly concave, and twice continuously differentiable, so that the first derivative is positive and the second negative; u' and u'' denote the first and second derivatives of u, respectively. Problem (2.1) is a classical problem in physics, called an isoperimetric problem; it arises when we seek to minimize the energy in a hanging string of fixed length. We make the mathematics of this problem slightly easier if we replace the integral constraint $\int_0^\infty c_t \, dt \leq s_0$ by a differential equation that equates the rate of change of the remaining stock to the consumption rate, and add an inequality requiring the remaining stock to be nonnegative:

$$\max \int_0^\infty u(c_t)e^{-\delta t} \, dt \quad \text{s.t.} \quad s_t \geq 0 \quad \text{and} \quad \dot{s}_t = -c_t \qquad (2.2)$$

where of course $s_t = s_0 - \int_0^t c_f \, df$. A dot over a variable is always used to denote its time derivative. Problems (2.1) and (2.2) are fully equivalent.

We solve this by introducing the Hamiltonian

$$H = u(c_t)e^{-\delta t} - \lambda_t e^{-\delta t} c_t \qquad (2.3)$$

where λ_t is an adjoint variable or shadow price, and then maximizing the Hamiltonian H with respect to c_t, giving

$$u'(c_t) = \lambda_t \, \forall t \qquad c_t > 0 \qquad (2.4)$$

In addition, the first-order conditions for a solution to (2.1) or (2.2) require that the present value of the adjoint variable or shadow price change over time at a rate given by the negative of the derivative of the Hamiltonian with respect to the stock:[1]

$$\frac{d}{dt} \lambda_t e^{-\delta t} = -\frac{\partial H}{\partial s_t}$$

so that

$$\dot{\lambda}_t - \delta \lambda_t = 0, \quad \text{i.e.,} \quad \lambda_t = \lambda_0 e^{\delta t} \qquad (2.5)$$

---

[1]These are standard techniques from control theory or calculus of variations. An intuitive exposition of these techniques can be found in Heal [57] and a more comprehensive treatment in Seirstad and Sydsaeter [99], or in Kamien and Schwartz [67]. A collection of articles applying these techniques to resource management is in Heal [59].

Equation (2.5) is known as the Hotelling Rule. It tells us that the present value of the shadow price of the resource has to be the same at all dates at which a positive amount is consumed. Of course, (2.4) tells us that the derivative of utility with respect to consumption, or marginal utility, must equal the shadow price, and so marginal utility also must grow at the discount rate and be constant in present value terms. This aspect of the result is very intuitive: consumption must be spread over the possible dates so that its incremental contribution to utility, in present value terms (i.e., its contribution to the maximand) is the same at all dates. This is the usual result that we spread a fixed factor between uses (dates, in this case) so that the incremental contribution that it makes is the same in all. When this condition is satisfied, no small variation in the time pattern of consumption will lead to an increase in the maximand.

What are the implications of this for consumption paths? To start with, consider a simple case. Let $u(c_t) = \log c_t$. Then (2.4) and (2.5) imply that

$$c_t = c_0 e^{-\delta t}$$

that is, consumption falls exponentially at the discount rate. Nothing is conserved or sustained forever, and the present and future are treated very unequally. The ratio c_t/c_0 of initial consumption c_0 to consumption at date t, c_t, decreases exponentially with time: $c_t/c_0 = e^{-\delta t}$. The inequality between generations increases exponentially over time.

In the general case, we have from (2.4) and (2.5) that

$$u''(c_t)\dot{c}_t = \delta u'(c_t)$$

so that

$$\frac{\dot{c}}{c} = -\frac{\delta}{\eta} \tag{2.6}$$

where $\eta = -c_t u''(c_t)/u'(c_t) > 0$ and is the elasticity of marginal utility of consumption; it is also a measure of risk aversion and of the curvature of the function $u(c)$. So we have the following general result:

PROPOSITION 1 If the utility function u has a constant elasticity of marginal utility, then consumption on an optimal path that solves (2.1) falls over time at a rate that is linear in the discount rate, with the constant of proportionality being the inverse of the elasticity of marginal utility: $c_t = c_0 e^{-(\delta/\eta)t}$.

Again, we have inequality in the treatment of present and future; this can be reduced by reducing the discount rate to give a flatter consumption profile, but the ratio of present to future consumption still grows exponentially. And setting the discount rate equal to zero is not a solution to the problem, for in that case (2.6) tells us that consumption should be constant over time, and the only feasible constant consumption level is zero. For $\delta = 0$, problem (2.1) has no solution.[2] This is an example of the unsettling paradoxes to which a zero discount rate gives rise, as mentioned in the discussion of discounted utilitarianism in Chapter 1.

2.1 Declining Discount Rate

In later chapters we find reasons to study optimal resource use problems in which the discount rate falls over time. The precise motivation for this is discussed in detail in chapters 5 and 7. For the moment we just put on the record some aspects of a solution for a particularly interesting example of a nonconstant discount rate. Consider the following problem:

$$\max \int_0^\infty u(c_t) e^{-\delta \log t}\, dt \quad \text{s.t.} \quad s_t \geq 0 \quad \text{and} \quad \dot{s}_t = -c_t \qquad (2.7)$$

By comparison with the original formulation (2.2), we have replaced δt by $\delta \log t$. We have a discount rate that is constant if instead of time t we work with its logarithm $\log t$; in other words, it is constant if instead of time we work with a logarithmic transformation of the time axis. This implies that the discount rate is declining with respect to the usual measure of time. In fact, in this case the rate of change of the discount factor $e^{-\delta \log t}$, which is the discount rate, is $-\delta/t$. So the effective discount rate declines asymptotically to zero. What we have here is a nonlinear transformation of the time axis, to a logarithmic scale. The reason why such a transformation might be interesting comes in chapters 5 and 7.

In this case, the same arguments as used in the previous section show that the shadow price now rises with a power of time, not exponentially

---

[2]For a detailed discussion of this case, see Heal [57] and Dasgupta and Heal [41].

with time:

$$\lambda_t = \lambda_0 t^{\delta}$$

For the logarithmic utility function used as an illustration before, consumption falls according to the power of time:

$$c_t = c_0 t^{-\delta}$$

Note that although consumption still goes to zero on an optimal path—it has no alternative, given the finite and unaugmentable stock—it now goes much more slowly, as a polynomial in t declines much more slowly than an exponential. In other words, inequality between generations now grows only polynomially in time, rather than exponentially with time. Changing the form of the discount factor has therefore made a qualitative difference in the way in which consumption declines, although of course it has not avoided the decline.

2.2   Conclusion

To summarize, the model used in this chapter—the Hotelling model of optimal depletion of an exhaustible resource—leaves little room for any sensible discussions of sustainability. The set of possible paths is very limited: consumption must go to zero, and as consumption is the only source of welfare, the economy must ultimately collapse. In chapter 3 we see that a small change, acknowledging an explicit value for the resource stock, alters a lot. It makes the problem qualitatively different. We still work with an exhaustible resource, so that the set of feasible paths is unaltered, but the valuation of the remaining stock alters optimal use patterns radically and introduces real substance into the discussion of sustainability. Making the resource renewable, which is the theme of chapter 4, takes this process even further. Another strategy for making the model richer is to allow for the accumulation of capital, which can to some degree substitute for the resource. This is the approach that was taken by Dasgupta and Heal [40, 41], who showed that positive consumption levels may be sustained forever even with an exhaustible resource, provided that there is considerable scope for substitution of produced capital for the resource. This issue is taken up at length in chapters 9 and 10.

2.3   Hamiltonians and Adjoint Variables

This section reviews very briefly and intuitively some key concepts and results associated with dynamic optimization as it is applied in later chapters. A complete and rigorous treatment can be found in [99]. We start with the problem considered in this chapter, which is very simple and intuitive; then we consider a much more general setting.

In this chapter we were concerned with the problem

$$\max \int_0^\infty u(c_t)e^{-\delta t}\, dt \quad \text{s.t.} \quad \int_0^\infty c_t\, dt \le s_0$$

Our first step was to replace the constraint $\int_0^\infty c_t\, dt \le s_0$ by the fully equivalent condition that

$$s_t = s_0 - \int_0^t c_f\, df \quad \text{and} \quad s_t \ge 0\, \forall t$$

So now we want to

$$\max \int_0^\infty u(c_t)e^{-\delta t}\, dt \quad \text{s.t.} \quad s_t \ge 0 \quad \text{and} \quad \dot{s}_t = -c_t$$

What we have done here is convert an integral constraint to a differential equation. We could work with either, but most results on dynamic optimization are stated in terms of problems where the constraint is in the form of a differential equation, so this is the more convenient form to work with. In this problem, the variable c_t is called the control variable; it is the variable with respect to whose value we are to optimize, the variable we have to choose at each point in time. The stock s_t is called the state variable; it defines the state of the system and its evolution from a given initial value is determined by the time path of the control variable.

The approach to solving such problems has much in common with the use of Lagrangians in solving static constrained maximization problems. In this one takes a constrained optimization problem and notes that its solution is the unconstrained optimum of a new function, the Lagrangian. This is formed by adding the constraints, multiplied by adjoint variables or shadow prices, to the objective. In the Hamiltonian approach to dynamic optimization we follow the same general route: the Hamiltonian is a function formed by adding to the function whose integral is the maximand, the right-hand side of the differential equation constraint multiplied by an adjoint variable or shadow price. So for the problem just set out, the Hamiltonian is

$$H = u(c_t)e^{-\delta t} - \lambda_t e^{-\delta t}c_t$$

where $\lambda_t e^{-\delta t}$ is the present value shadow price or adjoint variable: λ_t is the current value adjoint variable. Now we note that a necessary condition for the time path of c_t to be optimal in the overall problem is that at each date c should be chosen to give an unconstrained maximum of the function H. So far, what we are doing parallels exactly the standard procedures for solving static constrained maximization problems. Maximizing H with respect to c_t gives

$$u'(c_t) = \lambda_t \forall t \qquad c_t > 0$$

as a first-order condition. As we have already noted, this is a very intuitive condition, stating that whenever consumption of the resource is positive, the increase in utility from an increment of consumption should equal the shadow price or social valuation of the resource. Now, however, in contrast to the standard static optimization problem, we have a consumption level and shadow price at each point in time and must decide how these move over time. The second main result of the Hamiltonian approach tells us how the shadow price changes over time: it states that the rate of change of the present value of the shadow price must equal the negative of the derivative of the Hamiltonian with respect to the state variable. Formally,

$$\frac{d(\lambda_t e^{-\delta t})}{dt} = -\frac{\partial H}{\partial t}$$

In the context of the problem of this chapter, this rule implies

$$\frac{d\lambda}{dt} - \delta\lambda = 0 \quad \text{or} \quad \lambda_t = \lambda_0 e^{\delta t}$$

So the present value shadow price is constant, implying quite naturally that the present value of the contribution to utility of an increment of consumption should be the same at all dates. This fact, together with the first-order condition equating the derivative of consumption to the shadow price, allows us to characterize a solution to the optimization problem.

Next we look at a more complex problem to see how the dynamic optimization techniques apply in a general framework. The model we use for this is one that we use again in chapter 11 in our discussion of national income accounting. Let the vector $c_t \in \mathcal{R}^m$ be a vector of flows of goods consumed and giving utility at time t, and $s_t \in \mathcal{R}^n$ be a vector of stocks at time t, also possibly, but not necessarily, sources of utility. Each stock $s_{i,t}$, $i = 1, \ldots, n$, changes over time in a way that depends on the values of all stocks and of all flows:

$$\dot{s}_{i,t} = d_i(c_t, s_t), \qquad i = 1, \ldots, n \tag{2.8}$$

The economy's objective is to maximize the discounted integral of utilities, which may depend on stocks and flows:

$$\max \int_0^{\infty} u(c_t, s_t) e^{-\delta t} \, dt \tag{2.9}$$

subject to the rate-of-change equations (2.8) for the stocks. The utility function u is assumed to be strictly concave and the reproduction functions $d_i(c_t, s_t)$ are assumed to be concave. In this problem the control variables are as before the consumption levels c_t, and the state variables are the stocks s_t.

To solve this problem we construct a Hamiltonian that takes the form

$$H_t = u(c_t, s_t) e^{-\delta t} + \sum_{i=1}^{n} \lambda_{i,t} e^{-\delta t} d_i(c_t, s_t) \tag{2.10}$$

where the $\lambda_{i,t}$ are the current value shadow prices of the stocks $s_{i,t}$. The first-order conditions for maximizing the Hamiltonian with respect to the control variables $c_{j,t}$ can be summarized as

$$\frac{\partial u(c_t, s_t)}{\partial c_j} = -\sum_{i=1}^{n} \lambda_{i,t} \frac{\partial d_i(c_t, s_t)}{\partial c_j} \tag{2.11}$$

The condition that the rate of change of the present value shadow price equals the negative of the derivative of the Hamiltonian with respect to the associated state variable implies that

$$\dot{\lambda}_{i,t} - \delta\lambda_{i,t} = -\frac{\partial u(c_t, s_t)}{\partial s_i} - \sum_{k=1}^{n} \lambda_{k,t} \frac{\partial d_k(c_t, s_t)}{\partial s_i} \qquad (2.12)$$

Taken together, conditions (2.11) and (2.12) enable us to characterize a solution to the problem of maximizing (2.9) subject to (2.8). In later chapters, especially those related to national income and green accounting, we study the properties of the Hamiltonian (2.10) further, and see that it has an interesting connection with the Fisher–Lindahl–Hicks concept of income. For the time being, however, all that is needed is an understanding of how the Hamiltonian is used to derive the conditions necessary for optimality in dynamic optimization problems.

Chapter 3
Valuing a Depletable Stock

We have extracted as much as we can from the simplest pure depletion problem. The next step in the development of a satisfactory framework for analyzing sustainability is to make more of a concession to recognizing the different mechanisms through which environmental assets contribute to economic well-being. As a first move, we change the pure depletion problem by adding the remaining stock of the resource as an argument of the utility function. So now we recognize explicitly that the stock of the resource may be a source of value as well as the flow from it.

Examples of environmental resources for which this would be appropriate were mentioned in chapter 1. These include biodiversity, which in the sense of the range of species or some measure of their variation is a depletable asset: once it is reduced through extinction, it cannot be restored to its original value, at least on a time scale relevant to humanity. And clearly the stock of biodiversity is a source of many services, as mentioned in chapter 1.

Another example mentioned earlier is a forest, which yields a flow of wood for consumption as well as recreational facilities and carbon sequestration services, removing CO_2 from the atmosphere. Forests are usually renewable rather than depletable, but to a first approximation a tropical hardwood forest may be thought of as depletable. Other examples of depletable resources whose stocks provide value are a landscape, which can be farmed to yield a flow of output or enjoyed as a stock, or the atmosphere, which can be used to yield a flow of services as a sink for pollution or enjoyed as a stock of clean air.[1] To have a truly satisfactory framework for approaching sustainability, we need to

---

[1]This framework was introduced by Krautkraemer [71] and developed further by him in [72]. Many environmental resources that can be valued as stock as well as flows are renewable, such as forests. We reexamine the issues of concern in a renewable framework in chapter 4.

go beyond incorporating the services of the stock of natural assets to investigate objectives other than the discounted sum of utilities; we begin this task in chapter 5.

The basic problem is now

$$\max \int_0^\infty u(c_t, s_t) e^{-\delta t} dt \quad \text{s.t.} \quad s_t \geq 0 \quad \text{and} \quad \dot{s}_t = -c_t \qquad (3.1)$$

where the only alteration from chapter 2 is in the inclusion of the stock as an argument of the utility function. This change leads to qualitatively different conclusions. Now, for example, it may be optimal to preserve some of the resource stock indefinitely, in contrast to the previous case. How much should be preserved is sensitive to the precise specification of the objective, and we investigate several alternatives.

3.1   Corner Solutions

Before going into the formal analysis, I need to dwell on a technical point that will prove important in understanding the types of solutions that can emerge in problem (3.1). This point concerns the behavior of the utility function $u(c_t, s_t)$ as c goes to zero. Clearly, as the resource is exhaustible, the only sustainable consumption level is $c = 0$. However, in this case unlike that in chapter 2, there is a possibility that an optimal solution will have $c = 0$ but $s > 0$ in the long run. Now, the maintenance of a positive stock forever cannot happen if either the utility of zero consumption is minus infinity or the derivative of utility with respect to consumption is infinite at zero consumption:

$$\lim_{c \to \infty} \frac{\partial u(c_t, s_t)}{\partial c} = \infty \quad \text{or} \quad u(0, s_t) = -\infty$$

We therefore assume that neither of these conditions holds:

$$\lim_{c \to \infty} \frac{\partial u(c_t, s_t)}{\partial c} < \infty \quad \text{and} \quad u(0, s_t) > -\infty \, \forall s \qquad (3.2)$$

What we are ruling out here is an infinite penalty on zero consumption levels. The reason that this point merits specific attention is that many utility functions used in economics do not satisfy this condition: examples are the log utility function used in chapter 2, $u(c) = \log c$, and the function with constant elasticity of marginal utility $u(c) = -c^{-\eta}$. Both place an infinite penalty on zero consumption. This property is often deliberately chosen; it generally ensures that consumption is never zero. This is a reasonable requirement to impose on an optimal consumption path if the consumption one is considering is the sole source of welfare or is essential to survival, as in the case of food.

In the cases of the environmental assets that I want to model, assets whose stocks (as opposed to depletion rates) are sources of services to human societies, a flow of consumption from harvesting is usually not essential to survival. A flow of timber from tropical forests is not essential for our survival; in fact, it is a luxury. The existence of large stocks of such forests is more likely to be essential. Likewise, the consumption of biodiversity, in the sense of loss of genetic diversity, is not essential to us, whereas the maintenance of a stock of biodiversity may be. Again, the erosion of soil is not essential, whereas the maintenance of a stock is.

For these reasons I assume that the utility function does not impose an infinite penalty on zero consumption levels, as indicated in (3.2). This makes it possible, although not inevitable, for the model to have solutions on which $c = 0$ but $s > 0$ in the long run, a reasonable outcome if one thinks of commodities such as biodiversity, soils, or tropical hardwoods. There is undoubtedly some loss of generality here, but it seems acceptable given the focus of the study.

3.2   Optimal Paths

In the case of problem (3.1) the Hamiltonian is

$$H = u(c_t, s_t)e^{-\delta t} - \lambda_t c_t e^{-\delta t}$$

and maximization with respect to consumption c_t gives the same result as before, namely that the derivative of utility with respect to consumption must be greater than or equal to the shadow price of the resource:

$$u_c(c_t, s_t) \le \lambda_t, = \lambda_t \quad \text{if} \quad c_t > 0$$

where $u_c \equiv \partial u(c, s)/\partial c$, etc. However, the condition describing the movement of the shadow price over time is different, and now a solution has to satisfy

$$\dot{\lambda}_t - \delta \lambda_t = -u_s(c_t, s_t)$$

Note that these equations reduce to the earlier ones (2.4) and (2.5) if $u_s = 0$, as was implicitly assumed in chapter 2.

For simplicity now consider the case when the utility function is additively separable:

$$u(c_t, s_t) = u_1(c_t) + u_2(s_t)$$

where as always the u functions are increasing, strictly concave, and twice continuously differentiable. The implications of separability are that the marginal utility of consumption is independent of the level of the remaining stock and vice versa; in other words, the valuation of each source of utility is independent of the level of the other. Then, letting a prime denote the derivative of a function of one variable with respect to its argument, the conditions for optimality become

$$u_1'(c_t) \le \lambda_t, \ = \lambda_t \quad \text{if} \quad c_t > 0$$
$$\dot{\lambda}_t - \delta\lambda_t = -u_2'(s_t) \tag{3.3}$$

In the previous case, the shadow price of the resource grew indefinitely; now, in contrast, there may be a solution at which c_t, s_t, and λ_t are constant. Note that if consumption is constant, it must of course be zero; this is the only feasible constant consumption. And note that if the shadow price is constant, then $\delta\lambda_t = u_2'(s_t)$. So at a stationary solution of the first-order conditions (3.3),

$$\delta \le \frac{u_2'(s^*)}{u_1'(0)}, \ = \quad \text{if} \quad c_t > 0 \tag{3.4}$$

where s^* is the constant value of the remaining resource stock.

This equation has a simple interpretation: it requires that the slope of an indifference curve in the s–c plane, the ratio of the marginal utilities of the stock and flow, equal or exceed the discount rate. If we rewrite it as $u_1'(0) = u_2'(s^*)/\delta$, we can see another interpretation. Consider postponing an increment of consumption Δc indefinitely. The loss of utility is the derivative of utility with respect to consumption times the drop in consumption, $u_1'(0)\Delta c$. The gain from an increased stock, which continues indefinitely, is the present value of the stream of incremental utilities accruing from an increased stock.

$$\int_0^\infty u_2'(s^*) \Delta c \exp(-\delta t) \, dt = \Delta c u_2'(s^*)/\delta$$

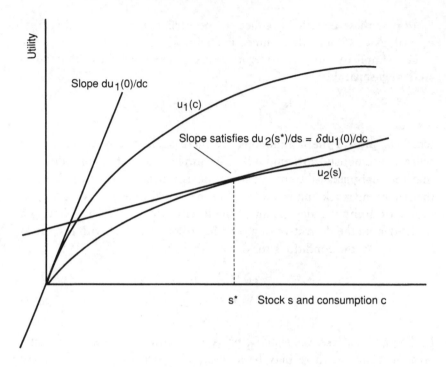

FIGURE 3.1 Determining the stationary stock of the environmental asset. The marginal utility of the stock equals the marginal utility of consumption at zero times the discount rate. The stock rises as the discount rate falls.

Equality of the incremental gains and losses implies (3.4), which can thus be interpreted as saying that the extra utility of an increment of consumption must equal the present value of the stream of incremental utility resulting from an increase in the stock. This is all very natural and straightforward.

The stationary configuration for this model is shown in figure 3.1; the constant level of the stock is one at which the derivative of utility with respect to the stock (the slope of the curve u_2) equals the slope of u_1, the utility-of-consumption function, at the origin, times the discount rate δ.

The dynamics of an optimal policy involve depleting the resource stock by consuming it until it is run down to s^*, and then stopping consumption and preserving the stock forever. The optimal policy is described by the two differential equations

$$u_1''(c_t)\dot{c}_t - \delta u_1'(c_t) = -u_2'(s_t)$$

$$\dot{s}_t = -c_t \tag{3.5}$$

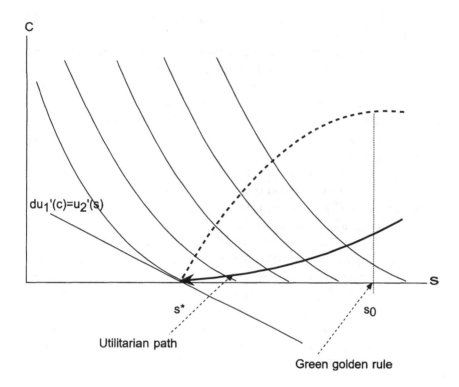

FIGURE 3.2 The dynamics of optimal paths when utility depends on the stock as well as the flow.

whose phase portrait is shown in figure 3.2. To understand this portrait, note that $\dot{s}_t$ is always nonpositive, so that in the horizontal dimension the system moves to the left or is stationary. $\dot{c}_t$ is zero on the curve $\delta u_1'(c_t) = u_2'(s_t)$, and is negative to the right of this and positive to the left. There is a stationary solution to the equations (3.5) at the point $c = 0, s = s^*$. It is straightforward to verify that this stationary solution is approached from initial points on a one-dimensional stable manifold, as shown in figure 3.2; this is shown by linearizing the systems (3.5) in a neighborhood of the stationary solution and observing that the matrix of the linearized system has real eigenvalues, one positive and one negative. The linearization is

$$\begin{bmatrix} \dfrac{u_2' u_1'''}{(u_1'')^2} + \dfrac{\delta([u_1'']^2 - u_1' u_1''')}{(u_1'')^2} & \dfrac{-u_2''}{u_1''} \\ -1 & 0 \end{bmatrix}$$

and it follows that the signs of the eigenvalues are opposite as the determinant is negative.

There is an important difference between the solution to the present problem and that to the classic Hotelling problem. In the present framework a positive stock may be preserved forever on an optimal path; exactly how much depends on the discount rate and on the utility functions, but this is a qualitative difference between the two problems. So the concept of sustainability seems to have some relevance in the context of this solution.

When will it be optimal in this context to preserve a positive stock level forever? This depends on the behavior of the utility function as the consumption level goes to zero. If the marginal utility of consumption goes to infinity as consumption goes to zero, as is often assumed, then equation (3.4) will have no solution and there will be no stationary state. This is the case with a Cobb–Douglas function, or with a separable utility function in which $u_1(c)$ has a constant elasticity of marginal utility. In any of these cases, the indifference curves in the c–s plane will not cross the horizontal axis, but will asymptote toward it. These are cases in which the flow from the resources is in some sense "essential" and cannot be allowed to fall to zero. It is not surprising that in these cases no stock will be preserved. We can summarize this as follows:

PROPOSITION 2 Consider an optimal solution to problem (3.1) when the utility function is additively separable, $u(c, s) = u_1(c) + u_2(s)$. A sufficient condition for this to involve the preservation of a positive stock forever is that the marginal utility of consumption at zero is finite, $u_1'(0) < \infty$, and that there exists a finite stock level s^*, the optimal stationary stock, such that $u_1'(0)\delta = u_2'(s^*)$. In this case, if the initial stock $s_0 > s^*$, then total consumption over time will equal $s_0 - s^*$; if $s_0 \leq s^*$, then consumption will always be zero and the entire stock will be conserved on an optimal path. On the other hand, if the marginal utility of consumption at $c = 0$ is infinite, then it will not be optimal to conserve any positive stock level indefinitely.

PROOF The only point in this proposition that needs to be proven is the statement that "if $s_0 \leq s^*$, then consumption will always be zero and the entire stock will be conserved on an optimal path." In figure 3.2, the claim we are making is that if $s_0 < s^*$, then the optimal solution is a point on the horizontal axis with $s_t = s_0 \forall t$. This follows from (3.4): these points on the horizontal axis with $s_0 < s*$ represent stationary solutions at which (3.4) is satisfied with inequality, which is acceptable as $c_t = 0$ at such points.

3.3 The Green Golden Rule

A second difference arising from the inclusion of the stock as a source of utility comes when we ask, "Which of all configurations of the economy gives the maximum sustainable utility level?" a question motivated by the golden rule of economic growth introduced in the 1960s by Phelps [90], and by our present interest in sustainability. In the Hotelling formulation, there is no interesting answer to this question; the only utility level maintainable forever is that associated with zero consumption. In the present model, however, the question is quite interesting, as many utility levels can be maintained forever. Clearly in the very long run, utility must be derived from the stock only, as no positive consumption can be maintained in the very long run. So the answer to the question must be "The utility level associated with the initial stock (the biggest stock ever) and zero consumption." Formally, in finding the maximum utility that can be sustained indefinitely we are maximizing $u(0, s)$ where $s \leq s_0$, and the solution is clearly to preserve the entire stock and never consume anything. This is the solution that we shall call the green golden rule, the path that gives the highest value of the long run level of utility:[2] it can be formalized as the solution to

$$\max_{\text{feasible paths}} \lim_{t \to \infty} u(c_t, s_t)$$

Formally,

PROPOSITION 3 The maximum sustainable utility level is attained by conserving the entire initial stock.

In the current model, though *not* in general, this is a solution that which leads to completely equal treatment of present and future. The same concept can be applied to the Hotelling problem, but the answer in that case is to consume nothing at any date. Although this does give the highest maintainable utility level, $u(0)$, in that problem, this illustrates

---

[2]This is formalized as the maximum limiting utility value. However, other formalizations are possible in principle, such as the maximum of the lim sup of the utility values. The differences between alternatives become significant only when positive limit sets of feasible trajectories may be limit cycles or other more complex attractors.

nicely a difficulty with this solution concept: in the Hotelling case this solution is clearly inefficient because, by consuming positive amounts, we can make some generations better off and none worse off. In general this solution concept is dynamically inefficient (it does not satisfy the first-order conditions for dynamics optimality), but in other cases this inefficiency is less obvious. We will see examples of this below. A key aspect of both of the problems analyzed so far has been the ability of the system to go directly to the green golden rule, without accumulating or decumulating any stock. As we shall see, this is not a general feature.

3.4   The Rawlsian Optimum

One of the alternative approaches to sustainability mentioned in the introduction was using the Rawlsian definition of justice between generations. In the case of the model used in this chapter, the green golden rule happens to be the path that is optimal in the Rawlsian sense (i.e., maximizes the welfare of the generation which is least well off). This point is seen rather easily. On any path that involves positive consumption, the utility level is nonincreasing over time. So the least well-off generation is the "last" generation; in fact, there is no "last" generation, so more accurately the lowest welfare level is the limiting welfare level. But this is maximized by the green golden rule, which maximizes the sustainable, and so the limiting, welfare level over all feasible paths. This coincidence of welfare criteria does not occur in all models, and in particular does not occur in the model of a renewable resource considered in chapter 4. In that model, the least well-off generation may be the first, not the last. Nevertheless, the importance of nonrenewable environmental resources, valued both as stocks and as flows, is probably sufficiently great that the present model has real relevance as an ideal type, so that the coincidence of criteria should not be viewed as a marginal result.

3.5   Nonseparable Utility Functions

We have worked so far with a utility function that is additively separable in the arguments, the flow of consumption c and the stock of the environmental asset s. Separability makes the algebra neat, but is not essential to the qualitative results. Here I just set out in outline the analysis for the more general case. The first-order conditions for optimality are now

$$\lambda = u_c(c, s)$$

$$\dot{\lambda} - \delta\lambda = -u_s(c, s)$$

where u_c is the derivative of $u(c, s)$ with respect to the argument c. A little manipulation yields

$$u_{cc}(c, s)\dot{c} + u_{cs}(c, s)\dot{s} = \delta u_c(c, s) - u_s(c, s)$$

At a point at which both c and s are constant,

$$\delta u_c(c, s) = u_s(c, s)$$

which is precisely the same characterization of a stationary solution as we obtained for the separable case, and can therefore be given the same intuitive interpretation. In this more general case, however, it is hard to characterize the locus of points at which $\dot{c} = 0$. In general, in the remainder of this book I shall often work with separable utility functions to simplify the exposition: nothing substantial will depend on this simplification.

3.6 Summary

Explicit recognition of the resource stock as a source of utility introduces some substance to the concept of sustainability. Now there is a purpose to conservation, and indeed full conservation emerges as the optimal policy for the objective of obtaining the highest sustainable utility level. On this interpretation, sustainability implies conservation, an identity many people will find appealing. However, we shall see that in more complex models this identity no longer holds. Interestingly, in the present framework, even discounted utilitarianism may recommend the conservation of a substantial stock ad infinitum.

Chapter 4
Renewable Resources

In this chapter we make a further addition to the structure of the model, this time in the specification of the constraints and of the dynamics of the resource. We now assume the resource to be renewable, that is, to have self-regenerating properties. The resource has a dynamic, a life, of its own. We model the interaction between this dynamic and the time path of its use by humans. Well-known examples that fall into this category are stocks of animals or fish, or forests. In fact, any ecosystem is of this type, and many of our most important natural resources are best seen as entire ecosystems rather than as individual species or subsystems. For example, soil is a renewable resource with a dynamic of its own, which interacts with the patterns of use by humans. Even for individual species such as whales or owls, one should ideally think of the validity and the dynamics of the entire ecosystem of which they are a part.

We shall see that the renewable nature of the resource makes a dramatic difference to the nature of optimal solutions. Now the future may actually be better treated than the present: if the initial resource stock is low, the optimal policy requires that consumption, stock, and utility all rise monotonically over time. The point is that because the resource is renewable, both stocks and flows can be built up over time provided that consumption is less than the rate of regeneration.

In this reformulation, the maximand remains exactly as before: primarily the discounted integral of utilities from consumption and from the existence of a stock, $\int_0^\infty u(c, s)e^{-\delta t}\, dt$, although as before some alternatives will be reviewed. However, the constraints are changed. We assume that the dynamics of the renewable resource are described by

$$\dot{s}_t = r(s_t) - c_t \tag{4.1}$$

(Note that this reduces to the previous formulation if the regeneration function $r[s]$ is identically zero.) Here r is the natural growth rate of the resource, assumed to depend only on its current stock. This describes its growth without human intervention. More complex models are of course possible, in which several such systems interact; a well-known example is the predator–prey system. In general, r is a concave function that attains a maximum at a finite value of s. This formulation has a long and classic history, which is reviewed in Dasgupta and Heal [41]. In the field of population biology, $r(s_t)$ is often taken to be quadratic, in which case an unexploited population (i.e., $c_t = 0 \forall t$) grows logistically. Here we assume that $r(0) = 0$, that there exists a positive stock level $\bar{s}$ at which $r(\bar{s}) = 0$, and that $r(s)$ is strictly concave and twice continuously differentiable.

Later chapters study a more complex formulation of the interaction between the dynamics of the economic and ecological systems; in particular, they include the possibility of capital accumulation as a part of the economic activities. Accumulated capital can in some degree substitute for the natural resource. Although this gives a more complex and realistic dynamic, the basic conclusions of the model are essentially unchanged.

Probably the weakest part of this specification is the ecological dynamic. As noted above and in earlier chapters, most ecosystems are considerably more complex than suggested by the adjustment equation (4.1). In most cases they consist of many linked elements, each with its own interacting dynamics. It is possible that under some conditions the simple representation used here can be thought of as an aggregate representation of the ecological system as a whole, with the variable s_t not the stock of an individual type but an aggregate measure such as biomass; this is a topic for further research. It is also true, fortunately, that the general qualitative conclusions we reach do not depend very sensitively on the precise specification of the ecological system.

The overall problem can now be specified as

$$\max \int_0^\infty u(c,s)e^{-\delta t}\, dt \quad \text{s.t.} \quad \dot{s}_t = r(s_t) - c_t, \qquad s_0 \text{ given} \quad (4.2)$$

The Hamiltonian in this case is

$$H = u(c_t, s_t)e^{-\delta t} + \lambda_t e^{-\delta t}[r(s_t) - c_t]$$

Maximization with respect to consumption gives as usual the equality of the marginal utility of consumption to the shadow price:

$$u_c(c_t, s_t) = \lambda_t$$

and the rate of change of the shadow price is determined by

$$\frac{d}{dt}(\lambda_t e^{-\delta t}) = -[u_s(c_t, s_t)e^{-\delta t} + \lambda_t e^{-\delta t} r'(s_t)]$$

To simplify matters we shall again take the utility function to be separable in c and s: $u(c, s) = u_1(c) + u_2(s)$, again taken to be strictly concave and twice differentiable. In this case a solution to the problem is characterized by

$$u_1'(c_t) = \lambda_t$$

$$\dot{s}_t = r(s_t) - c_t$$

$$\dot{\lambda}_t - \delta\lambda = -u_2'(s_t) - \lambda_t r'(s_t) \tag{4.3}$$

which in fact reduce to the equations of the solution to the previous problem if $r(s)$ is identically zero. In studying these equations, we first analyze their stationary solution, and then examine the dynamics of this system away from the stationary solution.

4.1 Stationary Solutions

At a stationary solution, s is constant so that $r(s_t) = c_t$; in addition the shadow price is constant so that

$$\delta u_1'(c_t) = u_2'(s_t) + u_1'(c_t)r'(s_t)$$

Hence,

PROPOSITION 4 A stationary solution to (4.3) satisfies

$$r(s_t) = c_t$$

$$\frac{u_2'(s_t)}{u_1'(c_t)} = \delta - r'(s_t) \tag{4.4}$$

The first equation in (4.4) just tells us that a stationary solution must lie on the curve on which consumption of the resource equals its renewal rate; this is obviously a prerequisite for a stationary stock. The second gives us a relationship between the slope of an indifference curve in the

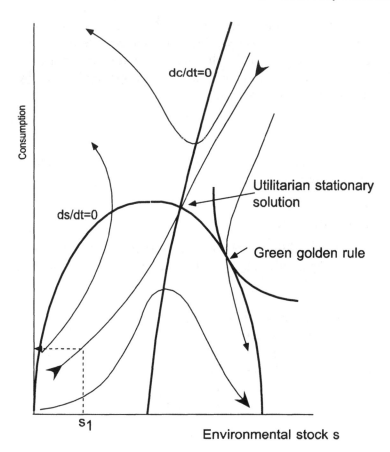

FIGURE 4.1 Dynamics of the utilitarian solution. The green golden rule—the highest sustainable utility level—is the point of tangency between an indifference curve and the growth curve.

c–s plane and the slope of the renewal function at a stationary solution; the indifference curve cuts the renewal function from above. Such a configuration is shown as the utilitarian stationary solution in figure 4.1. This reduces to the earlier result that the slope of an indifference curve should equal the discount rate if $r'(s) = 0 \forall s$, i.e., if the resource is nonrenewable.

There is a straightforward intuitive interpretation to the second equation in (4.4). This interpretation is exactly analogous to that given for the corresponding equation (3.4) in chapter 3. Consider reducing consumption by an amount Δc and increasing the stock by the same amount. The welfare loss is $\Delta c u_1'$; there is a gain from increasing the stock of $\Delta c u_2'$, which continues forever. But in addition the changed stock leads to a different stationary consumption level, in the amount $\Delta cr'$. Both of these effects—the changed stock and the changed stationary consumption—continue forever, so we have to take their present values by dividing by the discount rate. Hence at an optimum

$$\Delta cu_1' = \Delta c(u_2' + r'u_1')/\delta$$

which reduces to the second equation in (4.4).[1] So (4.4) is a very natural and intuitive characterization of optimality.

4.2 Dynamic Behavior

What are the dynamics of this system outside a stationary solution? These are also shown in figure 4.1. They are derived by noting the following facts:

1. Beneath the curve $r(s) = c$, s is rising as consumption is less than the growth of the resource.
2. Above the curve $r(s) = c$, s is falling as consumption is greater than the growth of the resource.
3. On the curve $r(s) = c$, s is constant.
4. From (4.3), the rate of change of c is given by

$$u_1''(c)\dot{c} = u_1'(c)[\delta - r'(s)] - u_2'(s)$$

The first term here is negative for small s and vice versa; the second is negative and large for small s and negative and small for large s. Hence c is rising for small s and vice versa; its rate of change is zero precisely when the rate of change of the shadow price is zero, which is on a line containing the stationary solution. The slope of this line is given by

$$\frac{\partial c}{\partial s} = \frac{u_1'r' + u_2''}{u_2''(\delta - r')}$$

The numerator is negative; the denominator is likewise if $\delta > r'$, in which case the slope of the $\dot{c} = 0$ line is positive at least in a neighborhood of the stationary solution.

---

[1] I am grateful to Jim Wilen for this interpretation.

5. By linearizing the system

$$u_1''(c)\dot{c} = u_1'(c)[\delta - r'(s)] - u_2'(s)$$

$$\dot{s}_t = r(s_t) - c_t$$

around the stationary solution, one can show that this solution is a saddle point. The matrix of the lincarized system is

$$\begin{bmatrix} r'(s) & -1 \\ -\dfrac{1}{u_1''}(u_1'r'' + u_2'') & [\delta - r'(s)] \end{bmatrix}$$

and the determinant of this is

$$r'(s)[\delta - r'(s)] - \frac{1}{u_1''}[u_1'r'' + u_2'']$$

which is negative for any stationary stock in excess of that giving the maximum rate of growth of the stock (i.e., for any stationary stock in excess of that giving the maximum sustainable yield).[2]

This shows that the utilitarian stationary solution (4.4) is a saddlepoint locally if it involves a stationary stock in excess of that giving the maximum sustainable yield. This is certainly the case for δ small enough. Hence the dynamics of paths satisfying the necessary conditions for optimality are as shown in figure 4.1, and we can establish the following result:

PROPOSITION 5 For small enough discount rates, optimal paths for problem (4.2) tend to the stationary solution (4.4). They do so along a path satisfying the first-order conditions (4.3), and follow one of the two branches of the stable path in figure 4.1 leading to the stationary solution. Given any initial value of the stock s_0, there is a corresponding value of c_0 that will place the system on one of the stable branches leading to the stationary solution. The position of

---

[2]It is also negative for small stocks for which $r' > \delta > 0$, and in other cases. To simplify the linearization I have taken the third derivative of u_2 to be zero, or at least small relative to the square of the second derivative.

the stationary solution depends on the discount rate, and moves to higher values of the stationary stock as this decreases. As $\delta \to 0$, the stationary solution tends to a point satisfying $u_2'/u_1' = r'$, which means in geometric terms that an indifference curve of $u(c, s)$ is tangent to the curve $c = r(s)$ given by the graph of the renewal function.

The renewable nature of the resource has clearly made a dramatic difference in the nature of optimal solutions. Now the future may actually be better treated than the present: if the initial resource stock is low, *the optimal policy requires that consumption, stock, and utility all rise monotonically over time.* The point is that because the resource is renewable, both stocks and flows can be built up over time provided that consumption is less than the rate of regeneration, i.e., the system is inside the curve given by the graph of the renewal function $r(s)$.

In practice, unfortunately, many renewable resources are being consumed at a rate much faster than their rates of regeneration; in terms of figure 4.1, the current consumption rate c_t is much greater than $r(s_t)$. So taking advantage of the regeneration possibilities of these resources would in many cases require sharp limitation of current consumption. Fisheries are a widely publicized example; another is tropical hardwoods and tropical forests in general. Soil is a more subtle example; there are processes that renew soil, so that even if it suffers a certain amount of erosion or depletion of its valuable components, it can be replaced. But typically human use of soils is depleting them at rates far faster than their replenishment rates.

4.3 The Green Golden Rule

We can use the renewable framework to ask a question that we asked before: what is the maximum sustainable utility level? There is a simple answer.

First, note that a sustainable utility level must be associated with a sustainable configuration of the economy, i.e., with sustainable values of consumption and of the stock. But these are precisely the values that satisfy the equation

$$c_t = r(s_t)$$

for these are the values that are feasible and at which the stock and the consumption levels are constant. Hence in figure 4.1, we are looking for values that lie on the curve $c_t = r(s_t)$. Of these values, we need the one that lies on the highest indifference curve of the utility function $u(c, s)$; this point of tangency is shown in the figure. At this point, the slope of an indifference curve equals that of the renewal function, so that the marginal rate of substitution between stock and flow equals the marginal rate of transformation along the curve $r(s)$. Hence,

PROPOSITION 6 The maximum sustainable utility level (the green golden rule) satisfies

$$\frac{u_2'(s_t)}{u_1'(c_t)} = -r'(s_t)$$

Recall from (4.4) that as the discount rate goes to zero, the stationary solution to the utilitarian case tends to such a point.

Note also that any path that approaches the tangency of an indifference curve with the reproduction function is optimal according to the criterion of achieving the maximum sustainable utility. In other words, this criterion of optimality determines only the limiting behavior of the economy; it does not determine how the limit is approached. This clearly is a weakness: of the many paths that approach the green golden rule, some will accumulate far more utility than others. One would like to know which of these is the best, or indeed whether there is such a best. We return to this later; it is the essence of Chichilnisky's concept of intertemporal equity, developed in chapter 5, that in the case of renewable resources it leads us to search for the best of the paths that approach the green golden rule. This search, as we shall see, is not easy.

4.4   Ecological Stability

An interesting fact is that the green golden rule, and also for low enough discount rates the utilitarian solution, require stocks of the resource that are greater than that giving the maximum sustainable yield, which is of course the stock at which the maximum of $r(s)$ occurs. This is important because only resource stocks greater than that giving the maximum sustainable yield are stable under the natural population dynamics of the resource (see Roughgarden and Smith [97]); they are ecologically stable. To see this, consider a fixed depletion rate d, so that the resource dynamics is just

$$\dot{s} = r(s) - d$$

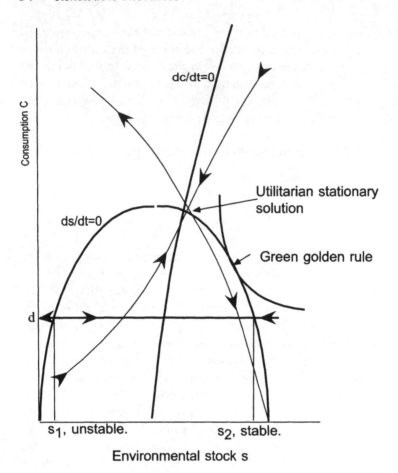

FIGURE 4.2 Dynamics of the renewable resource under a fixed depletion rate d.

For $d < \max\limits_{s} r(s)$, two values of s give stationary solutions to this equation, as shown in figure 4.2. Call the smaller s_1 and the larger s_2.
Clearly, when the stock exceeds the larger of these, s_2, it falls; i.e., for
$s > s_2$, $\dot{s} < 0$. Conversely, when the stock is between the two levels,
it rises; for $s_1 < s < s_2$, $\dot{s} > 0$. Finally, note that when the stock is
less than the smaller of the two values, it declines; for $s < s_1$, $\dot{s} < 0$,
as shown in figure 4.2. It follows that only the stock to the right
of the maximum sustainable yield is stable under the natural population adjustment process. High discount rates, and utilitarian optimal
policies when the stock of the resource is not an argument of the
utility function, will all give stationary stocks below the maximum

sustainable yield. The approaches pursued here, however, have the advantage of leading the economy to a configuration of the ecological subsystem that is stable under its normal dynamics. In practical terms this is important; it means that we are not trying to steer the resource stock toward a configuration at which it is inherently unstable (see [97]).

4.5   The Rawlsian Solution

In the nonrenewable context, we noted the coincidence of the Rawlsian optimum with the green golden rule. In the present case things do not always fit together so neatly. Consider the initial stock level s_1 in figure 4.1: the utilitarian optimum from this is to follow the path that leads to the saddle point. In this case, as noted, consumption, stock, and utility are all increasing. So the generation that is least well off is the first generation, not the last, as it was in the nonrenewable case. What is the Rawlsian solution in the present model, with initial stock s_1? It is easy to verify that this involves setting $c = r(s_1)$ forever; this gives a constant utility level, and gives the highest utility level for the first generation compatible with subsequent levels being no lower. This remains true for any initial stock no greater than that associated with the green golden rule: for larger initial stocks, the green golden rule is a Rawlsian optimum and in this case we do still have the coincidence noted in the previous chapter. Formally,

PROPOSITION 7 For an initial resource stock s_1 less than or equal to that associated with the green golden rule, the Rawlsian optimum involves setting $c = r(s_1)$ forever. For s_1 greater than the green golden rule stock, the green golden rule is a Rawlsian optimum.

4.6   Conclusions

This chapter and its two predecessors have reviewed some alternative optimal resource management policies for exhaustible and renewable resources. In the simplest of the models reviewed, the classical Hotelling model, there was no room for the concerns underlying sustainability. But once we introduced a concern for the extant stock as a source of economic benefits, this changed: some degree of preservation became an optimal policy under a wide range of circumstances. Making the resource renewable changed matters further. In this case any concept of optimality requires the preservation of a positive stock level, indeed for low enough discount rates a stock level greater than that giving the maximum sustainable yield. And in this case utilitarian optimal policies typically involve increasing resource stocks and increasing consumption levels over time. For initial resource stocks less than that giving the maximum sustainable yield, the ones likely in practice, the optimum policies start from levels of consumption that are less than the reproduction rate and are therefore sustainable in an intuitive sense. In other words, for most reasonable estimates of initial conditions, the optimal policies require that resources be managed so as to accumulate over time. Utilitarian optima are in some loose and intuitive sense sustainable in these models.

In chapter 5 we look in more detail at alternative concepts of optimality with respect to the management of a dynamic economy, as part of an effort to develop a more complete understanding of possible interpretations of sustainability.

Part II
A Broader Perspective

Chapter 5
Alternatives to Utilitarianism

Chapters 2, 3, and 4 completed the first step of this book. They reviewed the consequences of recognizing the importance of the stock of an environmental asset as a source of economic welfare, and introducing this as an argument of the utility function. They reviewed these consequences in the context of several criteria of optimality, namely, the discounted sum of utilities, the green golden rule, and the Rawlsian approach. The next step is to review systematically alternative concepts of intertemporal optimality, particularly concepts that place more weight on the future than does the conventional approach. This chapter is the introduction to that second step: it analyzes in detail alternative criteria of intertemporal optimality.

Economists have long sought a good alternative to discounted utilitarianism; as noted in chapter 1, the founders of growth theory, Ramsey [91] and Harrod [53] made notable and scathing comments about this framework. Less well known is the much earlier comment of Ramsey's predecessor at Cambridge, Sidgwick, one of the earliest exponents of utilitarianism, who commented that

> It seems...clear that the time at which a man exists cannot affect the value of his happiness from a universal point of view; and that the interests of posterity must concern a Utilitarian as much as those of his contemporaries, except in so far as the effects of his actions on posterity—and even on the existence of human beings to be affected—must necessarily be more uncertain. [Sidgwick [101], p. 412, quoted in Dasgupta and Heal [41], p. 262.]

These concerns have not died away with the passage of time; as recently as 1991, *The Economist* opined, in the context of environmental policy, that, "There is something awkward about discounting benefits that arise a century hence. For, even at a modest discount rate, no investment will look worthwhile." [*The Economist,* March 23rd, 1991, p. 73].

It seems that discounted utilitarianism has dominated until now largely because of the lack of a convincing alternative, not because of universal conviction about its merits.

In earlier chapters we reviewed alternative formulations of the economic model (exhaustible or renewable resources, utility as a function of consumption only or of consumption and the stock) and saw how they interact with alternative objectives: the discounted sum of utilities, the maximum sustainable utility level, and Rawlsian objective. The choice of objective makes a big difference. Seeking to approach the maximum sustainable utility level in a model in which both consumption and the remaining stock are arguments of the utility function probably comes closest to prescribing a strongly environment-oriented policy. It leads, with exhaustible resources, to the preservation of the entire stock. With renewable resources it leads to the selection of a stock in excess of that which gives the maximum sustainable yield. But it clearly has limitations: to consume nothing and conserve everything, in the exhaustible case, is extreme. And although the green golden rule in the renewable case is less extreme, its appeal is very limited as a solution, for it only tells us where we should be in the very long run. It does not define a path to follow. One might of course ask for "the best" path that leads there (best in the sense of the discounted sum of utilities along it), but in general this involves imposing on the discounted utilitarian problem a constraint—that the solution tend to the green golden rule—that is not justifiable within the utilitarian framework. In chapter 7 we see that there are exceptions to this: the green golden rule may be a solution to a utilitarian problem under some circumstances.

This suggests that something is missing from our set of possible objectives. For example, it would be appealing to have a definition of optimality that has the sensitivity to the long run of the green golden rule, but evaluates an entire path and not just its limiting properties. This chapter addresses these issues, with a view to seeing what is possible and reasonable.[1] We review systematically the alternatives to discounted utilitarianism that are more future oriented, and thus potentially more compatible with the concerns underlying sustainability. An interesting possibility, due to Chichilnisky, does emerge, and in later chapters we compare it to the alternatives we have considered so far.

---

[1] For a review of alternative criteria, see also Asheim [6].

5.1 Formalizing Utilitarianism

We should note that although the founders of utilitarianism and its dynamic applications were unanimous in condemning discounting, a justification for discounted utilitarianism has been provided by Harsanyi, ironically based on non-utilitarian arguments. He argued, along lines similar to those later used by Rawls, that a society should evaluate a utility stream as if it were unsure which generation it would be, which time slot it would occupy, and so which element of the sequence would indicate its welfare. In this case, he argued, risk aversion on the part of society would imply the representation of society's preferences over income or consumption streams in a form similar to the discounted utilitarian formulation. This matter is reviewed at length in Dasgupta and Heal [41].

The classic formalization of the discounted utilitarian approach for infinite horizons was provided by Koopmans [70] (see also Dasgupta and Heal [41]), who showed that ranking utility sequences by taking their present value at a constant discount rate follows from two conditions on the structure of preferences, plus several technical conditions. The central conditions he called *independence* and *stationarity;* these two conditions lead to a present value of utilities at a constant discount rate. Loosely speaking, independence requires that the marginal social rate of substitution between the welfares of any two generations be independent of the welfare of any third generation. Stationarity requires that if two utility sequences are equal in the first period, removing this period and advancing the remaining utility levels does not change their ranking.

One can formalize these conditions as follows. Consider a reference utility sequence $\bar{u} = \{\bar{u}_1, \bar{u}_2, \bar{u}_3, \ldots\}$. Consider also two other sequences $u = \{u_1, u_2, u_3, \ldots\}$ and $\hat{u} = \{\hat{u}_1, \hat{u}_2, \hat{u}_3, \ldots\}$. Then independence can be captured by the following requirements:

- If $\{u_1, \bar{u}_2, \bar{u}_3, \ldots\} \geqslant \{\hat{u}_1, \bar{u}_2, \bar{u}_3, \ldots\}$ then this ranking does not depend on the reference sequence $\bar{u}$.
- If $\{u_1, u_2, \bar{u}_3, \bar{u}_4, \ldots\} \geqslant \{\hat{u}_1, \hat{u}_2, \bar{u}_3, \bar{u}_4 \ldots\}$ then this ranking does not depend on the reference sequence $\bar{u}$.
- If $\{\bar{u}_1, u_2, u_3, \ldots\} \geqslant \{\bar{u}_1, \hat{u}_2, \hat{u}_3, \ldots\}$ then this ranking does not depend on the reference sequence $\bar{u}$.

Each of these requirements is a statement that the rate of substitution between the welfare levels of various generations does not depend on the welfare levels of those not under consideration.

The stationarity assumption can be made precise as follows: consider the utility sequences $\{u_1^*, u_2, u_3, \ldots\}$ and $\{u_1^*, \widehat{u}_2, \widehat{u}_3, \ldots\}$. Then $\{u_1^*, u_2, u_3, u_4, u_5, \ldots\} \geqslant \{u_1^*, \widehat{u}_2, \widehat{u}_3, \widehat{u}_4, \widehat{u}_5, \ldots\}$ if and only if $\{u_2, u_3, u_4, u_5, \ldots\} \geqslant \{\widehat{u}_2, \widehat{u}_3, \widehat{u}_4, \widehat{u}_5, \ldots\}$. The deletion of a first element common to both sequences and the advancement of the remainder does not affect the ranking of the remaining sequences.

Koopmans establishes that these conditions of independence and stationarity imply that any continuous and Paretian welfare ranking of alternative infinite utility sequences can be represented as

$$\sum_{t=1}^{t=\infty} \psi(u_t)\delta^t \quad \text{for} \quad \delta < 1$$

where ψ is an increasing continuous function unique up to a positive linear transformation. Independence is a strong condition; it implies, for example, that the trade-off between what happens today and what happens five years hence is independent of events in the interim.

5.2 Empirical Evidence

Before reviewing a range of theoretical alternatives, we consider briefly the empirical evidence on how individuals evaluate the future relative to the present. There is interesting evidence that they use a framework different in certain salient respects from the standard discounted utilitarian approach. Of course, even if we have a clear picture of how individuals form their judgments about the relative weights of present and future, this does not necessarily have normative implications: we might still feel that relative to some appropriate set of ethical standards they give too little (or too much) weight to the future, and so are an imperfect guide to social policy. However, in a democratic society, individual attitudes toward the present–future trade-off presumably have some informative value about the appropriate social trade-off and have at least an element of normative significance.

There is a growing body of empirical evidence (see for example Lowenstein and Thaler [80], Lowenstein and Prelec [79], papers in the volume by Lowenstein and Elster [78], Thaler [106], and Cropper et al. [36]) that suggests that the discount rate that people apply to future projects depends on, and declines with, the futurity of the project. Over short periods, up to perhaps five years, they use discount rates that are higher even than many commercial rates—in the region of 15% or in some cases much more. For projects extending about ten years, the implied discount rates are closer to standard rates, perhaps 10%. As the horizon extends the implied discount rates drops, to the region of 5% for thirty to fifty years and down to about 2% for one hundred years. It also appears from the empirical evidence that the discount rate used by individuals, and the way in which it changes over time, depends on the magnitude of the change in income involved.

5.3 Logarithmic Discounting and the Weber–Fechner Law

This empirically identified behavior has been called hyperbolic discounting (see Lowenstein and Prelec [79] or Ainslie and Haslam [1]) and is consistent with a very general set of results from natural sciences that find that human responses to a change in a stimulus are nonlinear, and are inversely proportional to the existing level of the stimulus (as mentioned in [1]).

For example, the human response to a change in the intensity of a sound is inversely proportional to the initial sound level: the louder the sound initially, the less we respond to a given increase. The same is true of responses to an increase in light intensity. These are illustrations of the Weber–Fechner law, formalized by the statement that human response to a change in a stimulus is inversely proportional to the preexisting stimulus. In symbols,

$$\frac{dr}{ds} = \frac{K}{s} \quad \text{or, integrating,} \quad r = K \log s$$

where r is a response, s a stimulus, and K a constant.

The empirical results on discounting cited above suggest that something similar is happening in human responses to changes in the futurity of an event: a given change in futurity (e.g., postponement by one year) leads to a smaller response in terms of the decrease in weighting, the further the event already is in the future. This is quite natural: postponement by one year from next year to the year after is clearly different from postponement from fifty to fifty-one years hence. The former obviously represents a major change; the latter, a small one. If we accept that the human reaction to postponement of a payoff or cost by a given period of time is indeed inversely proportional to its initial distance in the future, this suggests that the Weber–Fechner law can be applied to responses to distance in time, as well as to sound and light intensity. The result is that the discount rate is inversely proportional to distance into the future. Another way of saying this is that we react to proportional rather than absolute increases in the time distance. Denote the discount factor at time t by $\Delta(t) \leq 1$, so that this represents the weight placed on benefits at date t relative to a weight of unity at time zero. In this case the discount rate $q(t)$ is minus the rate of change of this weight over time,[2] or $q(t) = -\dot{\Delta}(t)/\Delta(t)$. We can formalize the idea that a given increase in the number of years into the future has an impact on the weight given to the event which is inversely proportional to the initial distance in the future as

$$q(t) = -\frac{1}{\Delta}\frac{d\Delta}{dt} = -\frac{K}{t} \quad \text{or} \quad \Delta(t) = e^{-K \log t} = t^{-K}$$

for a positive constant K. Such a formulation has several attractive properties: the discount rate q goes to zero in the limit,[3] the discount factor $\Delta(t)$ also goes to zero, and the integral $\int_1^\infty \Delta(t)\,dt = \int_1^\infty e^{-K\log t}dt = \int_1^\infty t^{-K}\,dt$ is finite for K positive and greater than unity.[4]

A discount factor $\Delta(t) = e^{-K\log t}$ has the interesting interpretation noted above; the replacement of t by $\log t$ implies that we are measuring time differently, by equal proportional increments rather than by equal absolute increments. We react in the same way to a given percentage increase in the number of years hence of an event, rather than to a given absolute increase in its number of years hence. We shall call this logarithmic discounting; this is consistent with the approach taken in acoustics, for example, where in response to the Weber–Fechner law sound intensity is measured in decibels, which respond to the logarithm of the energy content of the sound waves, and not to energy content itself. In general, nonconstant discount rates can be interpreted as a nonlinear transformation of the time axis.

5.4   Zero Discount Rate

Ramsey's own approach to defining intertemporal optimality was to use a zero discount rate when balancing the future against the present. In his

---

[2]The inclusion of a minus sign is required because it is conventional to report the discount rate as a positive number, whereas the rate of change of the discount factor Δ is normally negative.

[3]This is called slow discounting by Harvey [56].

[4]The lower limit of integration in this example is one, not zero, as t^{-K} is ill defined for $t = 0$ and $K > 0$.

particular formulation of the problem of optimal growth, this worked. He assumed the utility function to be bounded above (by what he called the bliss level), and worked in a framework in which the optimal policy leads to utility levels that approach the bliss level. Rather than maximize an integral of utilities, he then minimized the integral of the difference between the actual utility level and the bliss level. This is an ingenious approach, but in general it is not satisfactory. It requires that along an optimal path utility approaches its upper bound fast enough that the integral of the difference between the current utility value and the bound is finite. Formally, if B is the least upper bound of the utility function, then for this approach to work it has to be the case that

$$\int_0^\infty [B - u(c_t^*, s_t^*)]\, dt < \infty$$

where (c_t^*, s_t^*) is the optimal path of consumption and the resource stock. In general this is not true. In an interesting variant on Ramsey's approach, Le Kama [75] uses the utility level at the green golden rule instead of a least upper bound on utilities, and shows that in some cases the integral of the difference converges along an optimal path:

$$\int_0^\infty [GGR - u(c_t^*, s_t^*)]\, dt < \infty$$

where GGR is the green golden rule utility level.

If instead one uses the undiscounted integral of utilities, rather than the integral of the difference between utility and its upper bound, then in many cases the integral $\int_0^\infty u(c)\, dt$ is infinite on a set of feasible paths, so that there is no way of comparing these paths: the ranking provided is incomplete. For example, in the depletable resource problem with the stock as an argument of the utility function, any path that maintains a positive stock forever will lead to an infinite utility integral. Therefore, all such paths give the same undiscounted integral of utilities, independently of the stock level. This is clearly not satisfactory.

In the pure depletion problem due to Hotelling and considered in chapter 2, there is a different problem with a zero discount rate: the integral of utilities may be finite, but nevertheless the problem has no solution because the set of attainable values of the utility integral is open (see Dasgupta and Heal [41] for a detailed analysis).

5.5   Overtaking

In an attempt to avoid the problems of zero discount rates and yet give equal weight to present and future, von Weizäcker [107] introduced the overtaking criterion:

DEFINITION 8 A path c^1 is said to weakly overtake a path c^2 if there exists a time T^* such that for all $T > T^*$, we have

$$\int_0^T u(c_t^1)\, dt \geq \int_0^T u(c_t^2)\, dt$$

c^1 is said to strictly overtake c^2 if the inequality is strict.

This is also an ingenious approach: it replaces infinite integrals by finite ones, and says that one path is better than another if from some date on, cumulative utility on that path is greater. This relationship can be checked even if both cumulative utility totals go to infinity as $T \to \infty$, so this approach does to some degree extend the applicability of an approach based on a zero discount rate.

Unfortunately, this way of ranking paths is again incomplete; it is easy to construct pairs of paths, say c_t^1 and c_t^2, such that c_t^1 does not overtake c_t^2 and vice versa. Such a situation is shown in figure 5.1, and a more detailed analysis is given in Dasgupta and Heal [41]. Note that paths whose utility sums move like the stepped path in figure 5.1 are not paradoxes but can arise as candidates for optimal paths. For example, the paper by Ryder and Heal [98] gives an example in which consumption may oscillate on an optimal path, which will give rise to a utility sum that will increase over time with fluctuations in its rate of change, as shown in figure 5.1. Subsequently many more authors have shown that oscillations can characterize paths that are optimal in the discounted utilitarian sense.

Another limitation of the overtaking criterion, noted by Lauwers [73] [74], is that despite having a zero discount rate, it is not neutral with respect to timing, but clearly displays impatience. An example given by Lauwers is $\{1, 0, 1, 0, 1, 0, \ldots\}$ versus $\{0, 1, 0, 1, 0, 1, \ldots\}$; the former weakly overtakes the latter, and the converse is not true, although each is a permutation of the other, and they differ only in that one is lagged a single period behind the other. Another example can reinforce this point. Consider the sequences $\{1, 1, 1, 1, \ldots\}$, $\{0, 1, 1, 1, 1, \ldots\}$, $\{0, 0, 1, 1, 1, 1, \ldots\}$,

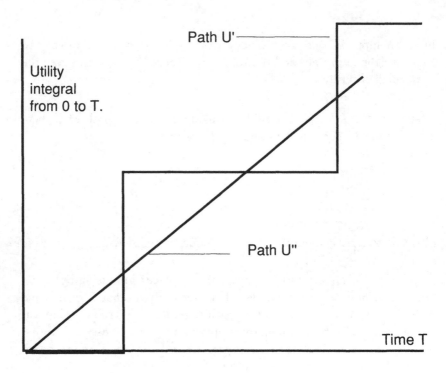

FIGURE 5.1   The overtaking criterion cannot rank paths u' and u''.

$\{0, 0, 0, 1, 1, 1, 1, \ldots\}$; clearly these have the same total consumption and differ only in that this is postponed further and further into the future. Equally clearly, the first strictly overtakes the second which strictly overtakes the third, etc. Once again, it is clear that the overtaking criterion with a zero discount rate is not neutral with respect to timing, but does display impatience.[5]

Finally, it is important to note a connection between the overtaking criterion and the limiting behavior of utility along a path. Intuitively, it is clear that if one path has a higher limiting utility value than another, then it overtakes the latter. This intuition is formalized in the following result. Consider two paths $U' = \{u'_1, u'_2, u'_3, \ldots\}$ and $U'' = \{u''_1, u''_2, u''_3, \ldots\}$ (we use the capital letter U to denote an infinite utility sequence: $U = \{u_1, u_2, u_3, \ldots\}$) and let both paths have limits: $\lim_{t \to \infty} u'_t = u'^*$ and $\lim_{t \to \infty} u''_t = u''^*$. Then:

---

[5]I am grateful to Yuliy Barishnikov for this example.

PROPOSITION 9 (1) A path with a higher limiting utility value always strictly overtakes one with a lower limiting utility value. Formally, the path U' strictly overtakes the path U'' if $\lim_{t \to \infty} u'_t = \overset{*}{u}{}' > \lim_{t \to \infty} u''_t = \overset{*}{u}{}''$. (2) If the path U' strictly overtakes the path U'', then its limiting utility value is not lower, i.e., $\lim_{t \to \infty} u'_t = \overset{*}{u}{}' \geq \lim_{t \to \infty} u''_t = \overset{*}{u}{}''$.

PROOF The proof of part (1) is very simple. Let $\overset{*}{u}{}' - \overset{*}{u}{}'' = \epsilon > 0$. Then there exists a time T_ϵ that depends on ϵ such that

$$\forall t > T_\epsilon, \left\| u'_t - \overset{*}{u}{}' \right\| < \frac{\epsilon}{4} \quad \text{and} \quad \left\| u''_t - \overset{*}{u}{}'' \right\| < \frac{\epsilon}{4}.$$

In this case $u'_t - u''_t > \frac{\epsilon}{2} \forall t > T_\epsilon$. Hence for $T > T_\epsilon$, $\sum_0^T (u'_t - u''_t) \geq \sum_0^{T_\epsilon} \sum (u'_t - u''_t) + (T - T_\epsilon)\frac{\epsilon}{2}$. The term $(T - T_\epsilon)\frac{\epsilon}{2}$ goes to infinity as T increases. This proves part (1).

The proof of part (2) is as follows. Suppose in contradiction that $\lim_{t \to \infty} u'_t = \overset{*}{u}{}' < \lim_{t \to \infty} u''_t = \overset{*}{u}{}''$. Then $\overset{*}{u}{}'' - \overset{*}{u}{}' = \epsilon > 0$. By the argument of part (1) u'' strictly overtakes u', which contradicts the assumption that u' strictly overtakes u''.

In summary, when we are dealing with utility paths that have limits, which is generally the case in optimal growth problems, then the overtaking criterion ranks paths with different limits by their limiting utility values. To check whether one path overtakes another, if they have different limits, one need only inspect their limits. If two paths have the same limit (for example, both tend to the green golden rule), then this observation is of no value, yet the overtaking criterion may still rank them. So on the space of utility paths that have limits, the overtaking criterion in effect acts lexicographically: first it ranks paths by their limiting values, and if these are equal, it then ranks them by their partial utility sums. One could easily strengthen proposition 9 to address the case in which neither path has a limit but $\liminf u' > \limsup u''$. One final comment on the overtaking criterion: unlike any of the other criteria we shall consider, it gives an ordering that cannot be represented by a real-valued function. There is no numerical function defined on alternative paths such that the path that overtakes all others gives the highest value of this function. If we restrict attention to the space of paths with different limits, there is such a function: it is just the limiting payoff. This representation cannot be extended to the space of all paths. This leads naturally to the next criterion.

5.6 Limiting Payoffs

An approach to evaluating intertemporal payoff streams, which is widely used in the theory of repeated games, is to rank paths by their long-run average payoffs, defined as

$$\lim_{m,n \to \infty} \left[\frac{1}{n} \sum_{m}^{m+n} u(c_t, s_t) \right]$$

This expression takes the average payoff over the period from m to $m+n$, and then takes the limit of this as m and n increase without limit. A similar approach is to rank according to limiting payoffs,[6] as in the cases considered of the green golden rule:

$$\lim_{t \to \infty} u(c_t, s_t)$$

As we have noted, this is in effect what the overtaking criterion does for paths that have limits. Of course, not all payoff sequences have limits, so the approach based on long-run averages is more general. As we see later in this chapter, the problem with these payoffs is that they neglect the present and overemphasize the long run. They can also lead to the selection of inefficient paths; this point was already noted in the context of the pure depletion problem formulated by Hotelling, and in the context of the green golden rule with a renewable resource. A criterion focused on evaluating paths according to their limiting behavior has nothing to say about almost all the paths. It ranks as the same two paths whose utility levels differ in every period, provided that they have the same limits. For example, consider the two utility sequences

$$\frac{1}{2}, \frac{3}{4}, \frac{7}{8}, \frac{15}{16}, \frac{31}{32}, \ldots \quad \text{and} \quad 0, 0, 0, 0, \frac{1}{2}, \frac{3}{4}, \frac{7}{8}, \frac{15}{16}, \frac{31}{32}, \ldots$$

Both have the same limit: 1. So ranking them according to their limits will rank them equally for any finite number of zeros at the start. Clearly this is not satisfactory.

---

[6] As noted before, other functions of the limiting behavior of the sequence are possible, such as lim sup. For a discussion of limiting criteria, see Dutta [47].

5.7   Chichilnisky's Criterion

Chichilnisky [22] proposes that we replace the discounted integral of utilities, or the long-run utility level, by the following maximand:

$$\alpha \int_0^\infty u(c_t, s_t) \, \Delta(t) \, dt + (1 - \alpha) \lim_{t \to \infty} (c_t, s_t) \qquad (5.1)$$

In this, $\alpha \in (0, 1)$ and $\Delta(t)$ is any measure, i.e., $\int_0^\infty \Delta(t) \, dt = 1$ (it has to be countably additive),[7] and in particular it could be a conventional exponential discount factor: $\Delta(t) = e^{-\delta t}$. The term lim could also be replaced by an alternative function that depends only on the limiting behavior of utility over time, such as the long-run average. Intuitively, this second term reflects the *sustainable utility level* attained by a policy. (Technically, any purely finitely additive measure will do.[8]) So Chichilnisky is in effect recommending a mixture of the two main approaches used so far: a generalization of the discounted utilitarian approach (to allow for any countably additive measure instead of just exponential discount factors with constant discount rates), mixed with the approach that ranks paths according to their very long-run characteristics or sustainable utility levels. It is this latter approach on its own that has led us to look at green golden rule solutions.

Chichilnisky does not pull this result out of a hat: she has a very precise rationalization, and shows that if we accept certain rather reasonable axioms about the ranking of alternative utility paths, then they *must* be ranked according to (5.1). Chichilnisky adopts two main axioms, which require that the ranking of alternative consumption paths be sensitive both to what happens in the present and immediate future, and also to what happens in the very long run.

Sensitivity to the long-run future is defined as follows: given any pair of consumption paths U' and U'' (we use the capital letter U to denote an infinite stream of utility values: $U = \{u_1, u_2, u_3, \ldots\}$), we can find no date T (which may depend on U' and U'') such that the ranking is insensitive to changes in the paths after the date T. In other words, there is no date such that changes after that date do not matter, in the sense of affecting the ranking. If we could find such a date, the ranking would be in an

---

[7] A countably additive measure is one that gives to any *countable* union of disjoint sets a measure that is the sum of the measures of the individual sets.

[8] A finitely additive measure is one that gives to any *finite* union of disjoint sets a measure that is the sum of the measures of the individual sets.

obvious sense insensitive to the long-run future, i.e., to the future beyond T. Hence the definition of sensitivity to the long-run future.

If this condition fails, then we can find a date T such that whatever changes are made in the two paths beyond T, their ranking will not be reversed. As an example, consider two utility sequences U' and U'', with U' ranked above U''. Suppose that no matter how much we improve U'' and worsen U' after the date $T(U', U'')$, we will never be able to change their ranking: U' will always rank higher. Such a situation is in an obvious sense insensitive to the long-run future.

Sensitivity to the present is defined symmetrically, as follows: given any two paths U' and U'', we can find no date T (which may depend on U' and U'') such that whatever changes are made in the two paths U' and U'' before T, their ranking will not be reversed. In other words, there is no date such that what happens before that date does not affect the ranking. If we could find such a date, the ranking would be in an obvious sense insensitive to the present and near future, i.e., the period up to T.

If this condition is not satisfied, then we can find a date T such that whatever changes are made in the two paths before T, their ranking will not be reversed. To illustrate, consider two utility sequences U' and U'', with U'' ranked above U'. Suppose that no matter how much we improve U' and worsen U'' before the date $T(U', U'')$, we will never be able to change their ranking; U'' will always rank higher. Such a situation is in an obvious sense insensitive to the present and near future.

These conditions can be formalized with the aid of the following definitions, which parallel closely those in Chichilnisky.

DEFINITION 10 Consider a sequence of utility values $U = \{u_1, u_2, u_3, \ldots, u_t, \ldots\}$. We define the T-th cutoff ^TU as $^TU = \{u_1, u_2, u_3, \ldots, u_T, 0, 0, 0, \ldots\}$. It is the sequence that agrees with U up to and including date T and is zero thereafter.

DEFINITION 11 Consider a sequence of utility values $U = \{u_1, u_2, u_3, \ldots, u_t, \ldots\}$. We define the T-th tail U^T as $U^T = \{0, 0, 0, \ldots, u_{T+1}, u_{T+2}, u_{T+3}, \ldots\}$. It is the sequence that agrees with U from date $T + 1$ onward and is zero before $T + 1$.

Clearly, $U = {^TU} + U^T$; U can be decomposed into its T-th tail and cutoff. Chichilnisky requires that a ranking (denoted $\succ$) of alternative utility sequences satisfy the following properties:

AXIOM 1 The ranking must be sensitive to the long-run future: formally, ∀ sequences U', U'' with $U' > U''$, for any date $T \; \exists$ tails $\tilde{U}^T, \hat{U}^T$ such that if these are substituted for the tails of U', U'' then the ranking is reversed, i.e., $\exists \; \tilde{U}^T, \hat{U}^T : \{^T U', \tilde{U}^T\} < \{^T U'', \hat{U}^T\}$.

AXIOM 2 The ranking must be sensitive to the present and near future: formally, ∀ U', U'' with $U' > U''$, there exists a date $T(U', U'')$ and cutoffs $^T\tilde{U}, {}^T\hat{U}$ such that if these are substituted for the cutoffs of U', U'' then the ranking is reversed, i.e., $\exists \; \tilde{U}^T, \hat{U}^T : \{^T\tilde{U}, U'^T\} < \{^T\hat{U}, U''^T\}$.

AXIOM 3 Continuity: the total measure of welfare of a utility sequence varies continuously (in the sup norm) with changes in the sequence.

AXIOM 4 The Pareto condition, i.e., if a change in a utility stream makes one generation better off and no other worse off, then it is ranked higher than before.

AXIOM 5 Linearity: the total welfare measure is linear in the welfares of generations.

Chichilnisky makes the additional assumption that the utility function $u(c_t, s_t)$ is bounded above and below:

$$\exists b_1, b_2 : b_1 \leq u(c_t, s_t) \leq b_2 \; \forall \, c_t, s_t$$

The function $u(c_t, s_t)$ is assumed the same for all dates t, so that generations are assumed to be the same in the way in which they rank alternatives. As $u(c_t, s_t)$ reflects only rankings of alternative states, it is a purely ordinal function and there is no loss of generality in the boundedness assumption. In particular, we are at liberty to make any order-preserving transformation of this function, provided that it remains the same function for all dates. Given this assumption, Chichilnisky proves the following:

PROPOSITION 12 If the ranking of alternative utility sequences satisfies axioms 1 to 5, then it must be represented by the following function (5.1):

$$\alpha \int_0^\infty u(c_t, s_t) \, \Delta(t) \, dt + (1 - \alpha) \lim_{t \to \infty} u(c_t, s_t), \qquad 0 < \alpha < 1$$

where $0 < \alpha < 1$, $\Delta(t)$ is any countably additive measure and the term lim could be replaced by any other purely finitely additive measure. There is no other way of ranking paths that meets all of the axioms.

Note that the first term here, the integral of utilities against a measure, which is analogous to a discounted integral of utilities, is insensitive to the long-run future, or is a dictatorship of the present in Chichilnisky's terminology:

PROPOSITION 13 The term $\int_0^\infty u(c_t, s_t)\Delta(t)\,dt$ is a dictatorship of the present; that is, it is insensitive to the long-run future.

PROOF This is easy to establish. Consider two utility paths u^1 and u^2, and assume that the utility integral against the measure $f(t)$ on the path u^1 exceeds that on the path u^2 by an amount K:

$$\int_0^\infty u^1 \Delta(t)\,dt = \int_0^\infty u^2 \Delta(t)\,dt + K, \qquad K > 0.$$

We will improve u^2 from some time T on by increasing it to the maximum utility level,[9] say $\bar{u}$, and worsen u^1 from that date on by reducing it to the minimum utility level, say $\underline{u}$, and show that if T is large enough even this will not reverse the ranking of the two paths. Now just pick T sufficiently large that

$$K > \int_T^\infty \Delta(t)(\bar{u} - \underline{u})\,dt$$

We can always do this because $\int_0^\infty \Delta(t)\,dt = 1$, so by picking T large enough we can make the integral from that date on as small as we wish.

The second term in Chichilnisky's criterion is insensitive to the present, or is a dictatorship of the future in Chichilnisky's terminology: this is obvious because it depends only on the limiting behavior of the payoffs. Formally:

PROPOSITION 14 The term $\lim_{t \to \infty} u(c_t, s_t)$ is a dictatorship of the future; that is, it is insensitive to the present.

---

[9] Recall that utilities are bounded above and below.

PROOF The proof of this is obvious.

The interesting point here is that while the first term is a dictatorship of the present (i.e., is insensitive to the long-run) and the second is a dictatorship of the future (i.e., is insensitive to the present and near future), the two together are neither. These observations are very consistent with the intuition that discounted utilitarianism is biased to the present, whereas the green golden rule, which maximizes long-run utility, possibly errs in the opposite direction. A combination, however, has the merits of both and the defects of neither.

Although the derivation of this ranking requires technical arguments, there is in fact an intuitive explanation. Observe that the first element in this expression is just a discounted sum of utilities; this is precisely the standard approach favored by economists and questioned by environmentalists. However, the criterion as a whole is a generalization of this approach by the addition of a term that values the very long-run (or limiting) behavior of the economy. Of course, if only short periods of time are at stake, this will make no difference. Similarly, if the economy's limiting behavior is totally determined by technological and resource constraints, the presence of the additional term will have no impact. It counts only if the time horizon is long and several alternative behavior patterns are available over that period. So it is a generalization of the standard approach, and differs only in precisely the case in which we are dissatisfied with the discounted utilitarian approach: the valuation of the very long run. It relates to that approach as general relativity relates to Newtonian mechanics: it supplements it without replacing it in routine applications.

Another way of understanding this approach is to observe that we would probably like to weight consumption in all periods—present and future—equally. However, this is not possible: one cannot weight all elements of an arbitrarily long stream equally and positively, for if one did, the weights would sum to infinity. A natural response is therefore to concentrate some weight on the present $\left[\int_0^\infty u(c_t, s_t)\,\Delta(t)\,dt\right]$ and some on the future $\left[\lim_{t \to \infty} u(c_t, s_t)\right]$. This is precisely what the expression (5.1) does.

5.7.1 Comparison with Overtaking — Our earlier analysis of the overtaking criterion suggests that it has some features in common with Chichilnisky's criterion. They both give weight to limiting utility values, and they both base a ranking of utility streams on the properties of the entire stream, not just of its limiting properties. What are the differences?

One that we have already seen is that overtaking gives only a partial ordering of the set of possible utility sequences; Chichilnisky's criterion gives a complete ordering. More fundamentally, overtaking has some characteristics of a dictatorship of the future: recall proposition 9, which states that if two utility sequences have limits that are different, the one with the higher limit always overtakes the other. This implies immediately that if we have two utility sequences with limits, one greater than the other, then there exists a date such that no changes in the sequences before that date can alter the ranking: for such pairs of sequences, overtaking *is* a dictatorship of the future. In general, however, it is not a dictatorship of the future. This means that, like Chichilnisky's criterion, overtaking uses as information both the limiting behavior of a path and its finite utility sums, but it uses them differently: it uses them lexicographically, first checking to see whether two sequences have different limits and in that case ranking by the limit, and otherwise looking at finite utility sums. As an illustration of the differences that follow from this, consider two sequences with different limits, $U' = \{u'_1, u'_2, u'_3, \ldots\}$ and $U'' = \{u''_1, u''_2, u''_3, \ldots\}$ with $\lim_{t \to \infty} u'_t = \overset{*}{u}{}' > \lim_{t \to \infty} u''_t = \overset{*}{u}{}''$. Then U' is preferred to U'' according to the overtaking criterion. However, Chichilnisky's criterion ranks them as $\alpha \sum_0^\infty \delta^t u'_t + (1-\alpha) \overset{*}{u}{}' \overset{<}{\underset{>}{=}} \alpha \sum_0^\infty \delta^t u''_t + (1-\alpha) \overset{*}{u}{}''$, so that their ranking may be in the reverse order of their limits if the difference of their discounted utility sums is sufficiently large in the opposite direction ("sufficiently large" depends on α). Figure 5.2 illustrates such a case: it requires that the sequence with the smaller limit have a larger sum of utilities in the near future, which offsets the difference in long-run behavior. Chichilnisky's criterion therefore puts less weight on the very long run than does the overtaking criterion; it does permit differences over the near future to compensate for limiting differences.

5.7.2 Constancy of Discount Rates — Nothing in Chichilnisky's axioms requires that the discount factor $\Delta(t)$ be of the form $e^{-\delta t}$, giving a constant discount rate. To attain constancy of the discount rate, one has to invoke the set of axioms due to Koopmans [70] and introduced in section 5.1. Recall that those axioms require preferences to satisfy *independence* and *stationarity*; these two conditions lead to a discount

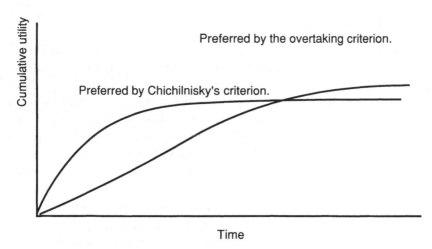

FIGURE 5.2 A case in which the overtaking criterion is defined and differs from Chichilnisky's.

factor implying a constant discount rate. Loosely speaking, independence requires that the marginal social rate of substitution between the welfares of any two generations be independent of the welfare of any third generation. Stationarity requires that if two utility sequences are equal in the first period, then removing this period and advancing the remaining utility levels does not change their ranking.

The central matter here is that the conditions implying a constant discount rate are distinct from those used in Chichilnisky's axiomatization. By imposing Koopmans's axioms in addition to Chichilnisky's, one would have a ranking of the form

$$\alpha \int_0^\infty u(c_t, s_t) e^{-\delta t}\, dt + (1 - \alpha) \lim_{t \to \infty} u(c_t, s_t), \qquad 0 < \alpha < 1$$

with an exponential discount factor and constant discount rate. Chichilnisky's formulation is compatible with constant discount rates, logarithmic discounting, or general nonconstant discounting. It is important to bear this in mind because, as we noted earlier in this chapter, there is empirical evidence that although people do discount the future, they do so not exponentially but at a declining discount rate. This matter of the constancy of the discount rate assumes a key role in the analysis later.

5.8 The Rawlsian Criterion

The Rawlsian criterion, choosing the path that maximizes the welfare of the least well-off generation, is often discussed in the context of sustainability (see for example Solow [104] and Asheim [4,8]). We considered its implications in earlier chapters in the context of models of the optimal use of both exhaustible and renewable resources. Where does it stand in weighing the present against the future? How does it compare in this important respect with the other criteria that have been reviewed?

On the key issue of present versus future, Rawls is ambivalent: Rawls is about rich versus poor, and will be pro-future if the future is poor, and vice versa. We have already seen this ambivalence. For the pure depletion model, it is strongly conservationist and strongly pro-future: it prescribes no consumption. But for a model with growth possibilities, as with the renewable resource model we have studied, Rawls's approach prescribes a policy that neglects the possibility of building up stocks and consumption over time at the cost of lower initial consumption. The reason is clear: Rawlsianism sanctions no sacrifice by the present for the future if this would make the present the poorest generation. Likewise, it sanctions no sacrifice by the future for the present if this would make the future the poorest period. It is neutral with respect to timing, but it will always rule out the most extreme exploitation of the future because of the strong egalitarianism inherent in the approach. In Chichilnisky's terms, Rawlsianism is insensitive not to both present or future, but to whatever time interval does not contain the poorest generation. It is sensitive only to the poorest generation, wherever that may be.

5.9 Discounting Utility or Consumption?

A distinction often made is between the rate at which utility is discounted (this is just the normal discount rate in the discounted utilitarian approach) and the rate at which consumption is discounted. The distinction is simple: both concepts arise within the discounted utilitarian framework, with utility discounted at a rate δ in the objective $\int_0^\infty u(c_t, s_t)e^{-\delta t}$.

Within this framework, one can ask the following question: suppose the economy follows a time path that is optimal for this criterion, and we consider adding an increment of consumption at some date t. What is the value of this increment in terms of its contribution to the objective function, and how does this value change as the date t is changed?

This contribution to the objective function is the value that should be assigned to an increment of consumption. The rate at which this contribution changes over time is the consumption rate of discount: it is the rate at which the weight of an increment of consumption, in terms of its contribution to the objective function, changes over time.

Clearly the value of an increment of consumption Δc at date t is

$$u_c(c_t, s_t)e^{-\delta t}\,\Delta c$$

and the rate at which this value changes with t is

$$\frac{1}{u_c(c_t, s_t)e^{-\delta t}}\frac{du_c(c_t, s_t)e^{-\delta t}}{dt}$$

which we can easily compute to be

$$-\delta - \eta_{c,c}\frac{\dot{c}}{c} - \eta_{c,s}\frac{\dot{s}}{s}$$

where $\eta_{c,c} = -cu_{c,c}/u_c > 0$ is the elasticity of the marginal utility of consumption with respect to the level of consumption and $\eta_{c,s} = -su_{c,s}/u_c > 0$ is the elasticity of the same quantity with respect to the level of the stock of the environmental asset. Consider for simplicity the case in which the utility function is additively separable, so that the cross-derivative is zero and $\eta_{c,s} = 0$. Then the consumption rate of discount is $\delta + \eta_{c,c}(\dot{c}/c)$. For a linear utility function, or for variations in the level of consumption small enough that a linear approximation to the utility function suffices, this reduces to δ, the utility rate of discount; the two concepts are the same. In this case we have a positive consumption rate of discount if and only if we have a positive utility rate of discount.

More generally, the two discount rates differ, and if consumption falls over time so that $(\dot{c}/c) < 0$, then the consumption discount rate may be negative, so that the weight given to an increment of consumption actually rises over time. In fact, if an economy is following an optimal path in the utilitarian sense, then the first-order conditions for optimality give us further information about the consumption discount rate.

One obvious and general proposition is the following. If the utility function has an elasticity of the marginal utility of consumption that is bounded, and if the utilitarian optimal path tends in the limit to a stationary solution, then *at this stationary solution the consumption discount rate is equal to the utility discount rate.* This is immediate from the expression $\delta + \eta_{c,c}(\dot{c}/c)$ for the consumption discount rate: if $\dot{c}$ is zero at a stationary solution, then this expression is just δ, the utility discount rate. So in general the utility and consumption discount rates converge in the limit along utilitarian optimal paths.

Consider some particular cases in more detail. First, in the pure depletion model of Hotelling considered in chapter 2, the first-order condition equation (2.6) is

$$\frac{\dot{c}}{c} = -\frac{\delta}{\eta}$$

so that along an optimal path the consumption discount rate is always zero, whatever the utility discount rate. In fact this is obvious: the first-order condition for optimality is just that the marginal contribution of an increment of consumption to the objective should be the same at all times, which is precisely that the consumption discount rate be zero. So *whatever the utility discount rate, and however uneven the distribution of consumption between generations, the consumption discount rate is zero*. This shows that a *zero consumption discount rate does not imply any degree of equality of consumption over time.* Note that in this case there is no stationary solution to the first-order conditions for optimality, so the consumption and utility discount rates do not converge.

In the case of the model of chapter 3, where the stock of the environmental asset enters into the utility function, the first-order conditions are

$$u_1'' \dot{c}_t - \delta u_1' = -u_2'$$
$$\dot{s}_t = -c_t$$

and we can rewrite the first equation as

$$-\eta \frac{\dot{c}_t}{c_t} - \delta = -\frac{u_2'}{u_1'}$$

so that the consumption discount rate is now the marginal rate of substitution between the stock and the flow, the slope of an indifference curve in the c–s plane. This is always positive, and by the analysis of chapter 3 goes in the limit to the utility discount rate δ. A similar analysis holds for the renewable resource model of chapter 4.

In this case the consumption rate of discount along a utilitarian optimal path is

$$-\frac{u'_2}{u'_1} - r'$$

which is generally positive and again goes to the utility discount rate δ in the limit. In both of these cases the first-order conditions admit a stationary solution so that the utility and consumption discount rates converge in the limit. A similar convergence holds for models with capital accumulation and production (see Heal [60] and chapters 9 and 10). In this case, the limiting value of the consumption discount rate is the marginal productivity of capital, whose limiting value is in turn the utility discount rate.

In summary, the consumption and utility discount rates are not independent concepts: when there is a stationary solution to the utilitarian problem, they are equal asymptotically along a utilitarian optimal path, and are always linked by the first-order conditions. Furthermore, the fact that the consumption discount rate is zero does not imply any degree of intergenerational equity, as the Hotelling example shows clearly. So having a zero consumption discount rate is not a solution to the ethical problem that led Ramsey and Harrod to decry the discounting of future utilities.

In principle we could define and measure the consumption discount rate along a path that is not a utilitarian optimum. We could consider the impact on a discounted utility sum of marginal variations about any reference time path of consumption, but the result would still depend on the utility discount rate chosen, and would be fundamentally arbitrary.

The central point is that we cannot avoid the need to choose a utility discount rate by focusing instead on consumption discount rates. Nor can we justify a positive utility discount rate by invoking an argument that the consumption discount rate may be quite different.

5.10   Final Comments

A criterion sometimes taken as defining, or at least as a key element of, sustainability, is that welfare levels be nondecreasing over time. Although it has some intuitive appeal, this criterion does not seem satisfactory. To see this point, note that in the case of pure depletion considered in chapters 2 and 3, there is only one path that satisfies this condition:

this is the path with zero consumption forever. On any other feasible path consumption and thus utility must fall. By contrast, in the case of a renewable resource, this condition is satisfied by any utilitarian path with a positive discount rate provided that the initial resource stock is less than the utilitarian stationary stock, and is satisfied by no utilitarian path if the initial stock exceeds this value. Yet in the renewable case, utilitarian optima are always in some loose sense sustainable, in that they involve maintaining a positive stock of the resource forever; indeed, they involve maintaining a stock in excess of that which gives the maximum sustainable yield. The condition that utility levels be non-decreasing is therefore unhelpful.

What this illustrates is part of a more general proposition: it is difficult to judge the appropriateness of an optimality criterion defining intertemporal welfare without some awareness of its implications. Of course, there are certain minimal requirements—logical consistency, completeness, representability—and certain candidates can be eliminated by reference to these requirements. Once we are over these basic hurdles, we have to investigate how the criteria perform and whether they lead to choices that are intuitively in accordance with our beliefs. In this process, we may be forced to clarify or even revise our beliefs: there can be a two-way iteration between formal optimality criteria and the value judgments to which we subscribe. The following chapters begin this process.

Chapter 6
Depletion Revisited

In this chapter and the next, we revisit the earlier models of pure depletion and of optimal use of renewable resources. We apply to them the ideas from chapter 5 and its perspective on alternative approaches to defining intertemporal optimality.

In Chichilnisky's criterion we now have a new way of ranking or comparing alternative consumption paths, which is a complete ordering that places on the future more weight than does discounted utilitarianism but less than maximizing asymptotic utility and seems to share some of the better characteristics of both. We will study the earlier models of optimal depletion in light of this criterion and see what it adds to an analysis of possible optimal paths and an exploration of the concept of sustainability.

Note that for the original Hotelling model of chapter 2, optima according to Chichilnisky's criterion and according to the discounted utilitarian criterion are the same. The presence of a term in the limiting utility value changes nothing in the analysis because only one limiting value is feasible, namely, that corresponding to a zero consumption level. The simplest framework in which this new element makes a difference is the model of depletion with both the flow and the stock of resources as sources of utility, exactly as set out in chapter 3. In this chapter we consider the impact of the new criterion in this context.

The problem of optimal resource use is therefore now[1]

$$\max \alpha \int_0^\infty u(c_t, s_t)e^{-\delta t}dt + (1 - \alpha)\lim_{t \to \infty} u(c_t, s_t), \qquad \alpha > 0, \tag{6.1}$$

$$\text{subject to} \quad \dot{s}_t = -c_t, \qquad s_t \geq 0 \, \forall t$$

[1]In this chapter we work with a formulation of Chichilnisky's criterion in which the discount rate is constant, although this is not implied by her axioms, as noted in chapter 5.

Unfortunately, there is no simple way of solving this problem. The usual rules from control theory or calculus of variations do not apply because of the presence of the term that depends only on the limiting behavior of the utility levels. We therefore have to use an approach tailored specifically to the problem.

6.1 Optimization with Chichilnisky's Criterion

We solve the optimization problem (6.1) by breaking it into two pieces, one corresponding to each part of the maximand. So we study the maximization of $\int_0^\infty u(c_t, s_t)e^{-\delta t}\, dt$ and then $\lim_{t \to \infty} u(c_t, s_t)$, subject to the constraint. After solving the separate problems, we put the solutions together. In other words, we solve the conventional variational problem of maximizing a discounted integral of utilities

$$\max \alpha \int_0^\infty u(c_t, s_t)e^{-\delta t}\, dt$$

$$\text{s.t.} \quad \dot{s}_t = -c_t, \qquad s_t \geq 0 \,\forall\, t$$

and then we solve the problem of maximizing the limiting utility value:

$$\max \lim_{t \to \infty} u(c_t, s_t) \quad \text{subject to} \quad \dot{s}_t = -c_t, \qquad s_t \geq 0 \,\forall\, t.$$

In fact we already know the solutions to both of these problems from chapter 3; the only question is how to combine them. We can combine them and solve the overall problem by constructing a new problem as follows: define a function $W(\beta s_0)$, which is the maximum present value utility that can be obtained if the initial stock of the resource is s_0, as before, but we are allowed to consume only a fraction β of this initial stock, withholding the rest as a stock that must be preserved forever. Formally this function is defined as follows:

$$W(\beta s_0) = \max \int_0^\infty u(c_t, s_t)e^{-\delta t}\, dt$$

(6.2)

$$\text{s.t.} \quad \int_0^\infty c_t\, dt \leq \beta s_0, \quad \text{initial stock} \quad s_0, \quad 0 \leq \beta \leq 1$$

In (6.2) we are optimizing subject to the constraint that the initial stock is s_0 and that not more than a fraction β of this initial stock s_0 is

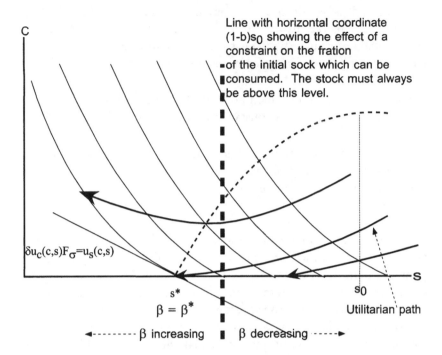

C

Line with horizontal coordinate $(1-b)s_0$ showing the effect of a constraint on the fration of the initial sock which can be consumed. The stock must always be above this level.

$\delta u_c(c,s)F_\sigma = u_s(c,s)$

s^*

$\beta = \beta^*$

s_0

Utilitarian path

S

◄ - - - - - - - β increasing ▮ β decreasing - - - - - ►

FIGURE 6.1   The effect of a constraint on the fraction of the initial stock that can be consumed.

consumed. Another way of stating this is that at least a fraction $(1 - \beta)$ must be preserved, so that $s_t \geq (1 - \beta)s_0 \ \forall t$ and the stock is at all times no less than $(1 - \beta)$ times the initial stock. This constraint is represented in figure 6.1, showing that the solution path must always lie to the right of the vertical line through $(1 - \beta)s_0$. As β increases, this line moves to the left and the constraint becomes weaker, and vice versa; if $\beta = 0$, then none of the stock can be consumed and the entire initial stock must be preserved.

Assume as before that the utility function is additively separable: $u(c, s) = u_1(c) + u_2(s)$. Let s^* be the stock level conserved forever in the discounted utilitarian solution (see figures 3.1 and 6.1). Define β^* by

$$\beta^* s_0 = s_0 - s^*$$

so that β^* is the fraction of the stock consumed along a utilitarian optimal path when there is no constraint on consumption. Intuitively one expects that as β increases and more of the stock may be consumed, this will increase, or at least not decrease, the maximum utility attainable, $W(\beta s_0)$. Furthermore, it also seems clear from figure 6.1 that once $\beta \geq \beta^*$, further increases will not change $W(\beta s_0)$ as the constraint $\int_0^\infty c_t \, dt \leq \beta s_0$ is not binding. The following lemma confirms this intuition.

LEMMA 15 The valuation function $W(\beta s_0)$ is non-decreasing as a function of β for $\beta < \beta^*$, and is constant as a function of β for $\beta \geq \beta^*$.

PROOF Increasing the value of β increases the set of permitted consumption paths in the problem (6.2). Consequently, the value of the maximand in this problem does not decrease, which proves the first part of the lemma.

Consider a solution to the optimization problem (6.2). The initial stock s_0 determines the horizontal coordinate of the initial point of an optimal path; the constraint $\int_0^\infty c_t \, dt \leq \beta s_0$ or equivalently $s_t \geq (1-\beta)s_0$ determines the upper bound to cumulative consumption along an optimal path and the lower bound to the terminal stock on this path.

First we study the case of $\beta \geq \beta^*$. In this range, β permits cumulative consumption levels above the maximum that will ever be used in the utilitarian problem. In this case the constraint $\int_0^\infty c_t \, dt \leq \beta s_0$ or equivalently $s_t \geq (1 - \beta)s_0$ has no impact and a solution to the problem (6.2) has the same cumulative consumption as the utilitarian solution. The solution to

$$\max \int_0^\infty u(c_t, s_t)e^{-\delta t} \, dt \quad \text{s.t.} \quad \int_0^\infty c_t \, dt \leq s_0$$

satisfies a tighter constraint than $\int_0^\infty c_t \, dt \leq \beta s_0$ and so is also optimal in the presence of this latter constraint. Hence the solution to the problem $\max \int_0^\infty u(c_t, s_t)e^{-\delta t} \, dt$ s.t. $\int_0^\infty c_t \, dt \leq s_0$ also solves the problem

$$\max \int_0^\infty u(c_t, s_t)e^{-\delta t} \, dt \quad \text{s.t.} \quad \int_0^\infty c_t \, dt \leq \beta s_0$$

so that an increase in β does not increase the maximand. In terms of figure 6.1, $\beta \geq \beta^*$ corresponds to the vertical line giving the boundary of the constraint set lying to the left of the utilitarian optimal stationary stock s^*.

Now consider $\beta < \beta^*$. The constraint $\int_0^\infty c_t\,dt \leq \beta s_0$ permits less consumption than is required on a utilitarian optimal path, or, equivalently, $s_t \geq (1 - \beta)s_0 > s^*$, as shown in figure 6.1. $W(\beta s_0)$ is less than the maximum attained the case of $\int_0^\infty c_t\,dt \leq s_0$. The Hamiltonian for (6.2) is

$$H = u(c_t, s_t)e^{-\delta t} + \lambda_t e^{-\delta t}(-c_t) + \mu[s_t - (1 - \beta)s_0]$$

where μ is the shadow price on the constraint $s_t \geq (1 - \beta)s_0$, which holds with complementary slackness, that is, $\mu[s_t - (1 - \beta)s_0] = 0$ and $\mu > 0$ if $s_t - (1 - \beta)s_0 = 0$. This gives rise to a dynamic system identical to that characterizing a solution to the utilitarian problem on the interior of the positive orthant, and on the boundary it has a stationary solution that satisfies

$$\frac{\partial u(c_t, s_t)}{\partial s_t} = \delta \frac{\partial u(c_t, s_t)}{\partial c_t} - \mu$$

for $\mu > 0$ and has a stationary value of the resource stock greater than the utilitarian solution s^*.

In this case the integral constraint is binding: a solution to (6.2) satisfies the same first-order conditions as the utilitarian solution but has a lower level of cumulative consumption and a higher terminal stock. The optimal path is a locally optimal path in figure 6.1, with initial consumption less than on the utilitarian optimal path. The value of the maximand along this path must be less than that along the utilitarian path with $\beta = 1$, and increases as β approaches 1. This completes the proof.

If we pick a consumption path to solve (6.2), then the total payoff from this path if we evaluate it according to the Chichilnisky criterion is given by the present value of utilities associated with this path, which is $\alpha W(\beta s_0)$, plus the limiting utility value, which is $(1 - \alpha) \lim_{t \to \infty} u(c_t, s_t)$. The expression for this depends on whether the amount of the initial stock made available for consumption, βs_0, is greater than or less than the amount consumed on a utilitarian optimal path, $\beta^* s_0$; that is, whether $\beta \gtrless \beta^*$. If $\beta < \beta^*$, the total payoff with the Chichilnisky criterion is

$$\alpha W(\beta s_0) + (1 - \alpha)u_2[(1 - \beta)s_0]$$

because in this case the constraint on cumulative consumption binds, the boundary of the constraint set in figure 6.1 lies above s^*, and the stock $(1 - \beta)s_0 > s^*$ is maintained forever. If $\beta \geq \beta^*$, the cumulative consumption constraint is not binding, and the limiting stock is that associated with the utilitarian optimum, s^*. This payoff according to the Chichilnisky criterion is now

$$\alpha W(\beta s_0) + (1 - \alpha)u_2(s^*)$$

The overall problem (6.1) can now be solved by picking β to maximize the sum of $\alpha W(\beta s_0)$ and either $(1 - \alpha)u_2[(1 - \beta)s_0]$ or $(1 - \alpha)u_2(s^*)$ as appropriate; of course, $(1 - \alpha)u_2[(1 - \beta)s_0] = (1 - \alpha)u_2(s^*)$ if $\beta = \beta^*$. The approach to solving this problem is illustrated in figure 6.2. From the previous lemma, we know that $\alpha W(\beta s_0)$ must be nondecreasing in β up to β^*, the point where βs_0 is equal to the optimal cumulative consumption level in the discounted utilitarian formulation, and that for β above the value β^*, $\alpha W(\beta s_0)$ must be constant because the possibility of consuming more than would be consumed in the unconstrained problem will clearly not increase the integral of discounted utilities. The constraint on consumption is not binding in this case.

Similarly, $(1 - \alpha)u_2[(1 - \beta)s_0]$ is nonincreasing in β to the point where $\beta = \beta^*$; for β greater than this the second term in the Chichilnisky criterion is $(1 - \alpha)u_2(s^*)$, and is not affected by a change in the constraints. So for $\beta \geq \beta^*$, the total payoff is constant as β varies and is at the level that solves the discounted utilitarian problem. The optimal choice of β is the value that maximizes the sum of the two curves in figure 6.2. Figure 6.2 shows this as a value less than β^*, the utilitarian optimum, and so implies a more conservation-oriented approach. We can summarize this discussion in the following proposition, which characterizes an optimal consumption pattern for the Chichilnisky criterion in the case of an exhaustible resource as modeled by problem 6.1.

PROPOSITION 16 Let $W(\beta s_0)$ be defined as in (6.2) above, let $[c_t(\beta)]$ be the time path of consumption that attains $W(\beta s_0)$, let s^* be the stock level conserved forever in the discounted utilitarian solution, and define β^* by $\beta^* s_0 = s_0 - s^*$. Let $\hat{\beta}$ be the value of β that maximizes

$$\alpha W(\beta s_0) + (1 - \alpha)u_2[(1 - \beta)s_0] \quad \text{if} \quad \beta \leq \beta^*$$

$$\alpha W(\beta s_0) + (1 - \alpha)u_2(s^*) \quad \text{otherwise}$$

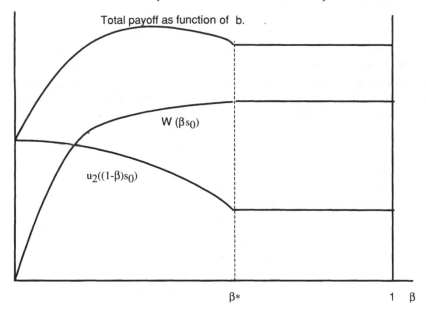

Total payoff as function of b.

$W(\beta s_0)$

$u_2((1-\beta)s_0)$

$\beta*$ 1 β

FIGURE 6.2   The total payoff as a function of β.

Then $[c_t(\hat{\beta})]$ attains a maximum of the Chichilnisky criterion

$$\alpha \int_0^\infty u(c_t, s_t)e^{-\delta t}\, dt + (1 - \alpha) \lim_{t \to \infty} u(c_t, s_t)$$

subject to the constraint $\dot{s}_t = -c_t$, $s_t \geq 0\,\forall\, t$.

A verbal interpretation of this proposition may be helpful. It says that given any Chichilnisky problem (6.1) there exists a utilitarian problem

$$\max \int_0^\infty u(c_t, s_t)e^{-\delta t}\, dt \quad \text{s.t.} \quad \int_0^\infty c_t\, dt \leq \hat{\beta} s_0,$$

initial stock s_0, where $\hat{\beta}$ is defined in proposition 16

whose solution is the same at that of the Chichilnisky problem. Clearly, the value of $\hat{\beta}$ in the equivalent utilitarian problem depends on that of α in the Chichilnisky problem; obviously, for $\alpha = 1$, $\hat{\beta} = 1$. Note that although the Chichilnisky problem has an equivalent utilitarian problem with the same solution, one cannot arrive at this problem by inspecting the Chichilnisky problem. There seems to be no way of computing

$\hat{\beta}$ other than going through the exercise in the previous proposition. Although the existence of an equivalent utilitarian proposition is not constructive in the sense of providing a quick way of solving problems with Chichilnisky's criterion of optimality, it does give some insight into the impact of adopting this criterion: it shows that what we are doing is effectively "freezing" a part of the initial stock of the resource and holding it in perpetuity for the benefit of all generations. This is intuitively consistent with the motivation underlying this criterion.

The dynamics of the solution are shown in figure 6.3. This corresponds to the configuration of figure 6.2. The path that is optimal according the Chichilnisky criterion leads to a long-run stock that is greater than the utilitarian one, but less than the green golden rule. In the next section we investigate whether this is generally true. The path starts from a lower consumption level, and runs consumption down to zero sooner. An important point about this path is that it satisfies the first-order conditions for optimality in the discounted utilitarian problem, so that it

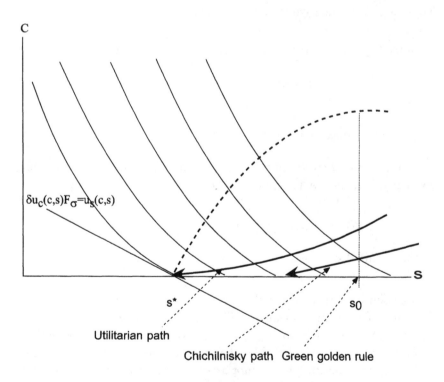

FIGURE 6.3 The dynamics of depletion paths optimal according to alternative optimality criteria.

is one of the paths in the discounted utilitarian phase portrait in figure 3.2. In fact this is a general proposition:

PROPOSITION 17 A path that is optimal with respect to the Chichilnisky criterion (5.1)

$$\alpha \int_0^\infty u(c_t, s_t)\,\Delta(t)\,dt \ + \ (1 \ - \ \alpha) \lim_{t \to \infty} u(c_t, s_t)$$

subject to certain constraints must always satisfy the conditions necessary for the maximization of the first term in (5.1), $\int_0^\infty u(c_t, s_t)\,\Delta(t)\,dt$, subject to the same constraints, on the interior of the positive cone.

PROOF Suppose that over the period in which $c_t > 0$, the optimal path in the Chichilnisky problem did not satisfy the first-order conditions for the utilitarian problem of maximizing $\int_0^\infty u(c_t, s_t)\,\Delta(t)\,dt$ subject to the appropriate constraint. Then it would be possible to rearrange the total consumption over this period in such a way as to increase the integral of discounted utilities without reducing the long-run utility level. In this case the policy could not have been optimal.

Note that this proposition does not apply only to the depletion problem with stock-dependent utility; it applies quite generally.

6.2 Conservation and the Chichilnisky Criterion

When does the Chichilnisky criterion lead to more conservation, to a larger long-run stock, than the discounted utilitarian framework with the same discount rate? This is the case shown in figures 6.2 and 6.3. It does so when the optimal choice of β is strictly less than β^*, that is, when the maximum of the sum of the curves in figure 6.2 is to the left of β^*. The following proposition gives conditions under which this occurs: these are simply that there should be a maximum in figure 6.2 in the interval $(0, \beta^*)$. Note that in figure 6.2, selecting the green golden rule, which requires that we maintain the entire initial stock intact forever, corresponds to selecting zero as the optimal value of β. Similarly, selecting $\beta = \beta^*$ corresponds to picking the path that maximizes the discounted integral of utilities.

PROPOSITION 18 The optimal path according to the Chichilnisky criterion involves a larger permanent stock than the utilitarian criterion if and only if there exists a $\beta \in (0, \beta^*)$ at which

$$\frac{\alpha}{1 - \alpha} = \frac{u'_2[(1 - \beta)s_0]}{W'(\beta s_0)}.$$

If

$$\frac{\alpha}{1 - \alpha} > \frac{u'_2[(1 - \beta)s_0]}{W'(\beta s_0)},$$

for all $\beta \in (0, \beta^*)$, then the two criteria lead to the preservation of the same stock.

PROOF The proof is immediate.

Figure 6.4 compares the optimal paths of s in this problem for the alternative optimality criteria we have examined.

By using an argument analogous to that following equation (3.4), we can characterize a stationary solution under the Chichilnisky criterion. Recall that for the discounted utilitarian criterion, the characterization is (3.4), which is

$$\frac{u'_2(s^*)}{u'_1(0)} = \delta$$

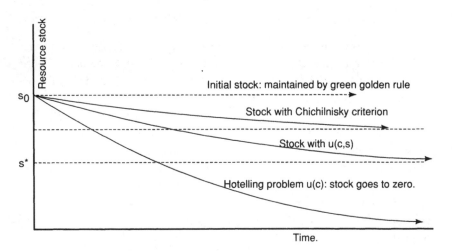

FIGURE 6.4 The time paths of the environmental stock under alternative optimality concepts.

which implies that the loss in utility of consumption from reducing consumption marginally must equal the present value of the gain from the consequent permanent increase in the stock. In the present context, we need an extension of this statement. The increase in the stock now contributes to two terms in the maximand, the term $\alpha \int_0^\infty u_2(s_t) \exp(-\delta t)\, dt$ and the term $(1 - \alpha) \lim_{t \to \infty} u_2(s_t)$. So now the loss in utility from a small reduction in consumption must just equal the present value of the increase in utility from the stock over the entire future plus the contribution of an increase in the permanent stock to the limiting utility value. Hence a stationary solution, now denoted $\hat{s}$, must satisfy

$$-\alpha u_1'(0) + \frac{\alpha u_2'(\hat{s})}{\delta} + (1 - \alpha)u_2'(\hat{s}) = 0 \tag{6.3}$$

which implies

$$\frac{u_2'(\hat{s})}{u_1'(0)} = \frac{\alpha\delta}{(1 - \alpha)\delta + \alpha} \tag{6.4}$$

This is clearly a generalization of (3.4), and reduces to this if $\alpha = 1$ and the limiting term is dropped from the maximand, giving the utilitarian case. For $\alpha < 1$, (6.4) implies a stationary capital stock greater than that implied by the utilitarian objective.

An important point about expression (6.3), which measures the consequences of a change in consumption, is the presence of the term $(1 - \alpha)u_2'(\hat{s})$, which is undiscounted despite representing benefits very far into the future. The presence of this is an obvious consequence of the form of the maximand, and is shown below to have important consequences for the measurement of national income, for the calculation of shadow prices, and thus for project evaluation.

6.3   Differences from Utilitarian Optima

How do solutions using the Chichilnisky criterion differ from discounted utilitarian solutions? In particular, could a solution with this criterion be replicated with a discounted utilitarian formulation with a lower discount rate? The answer is no.

Consider figure 6.3, and in particular the utilitarian optimal path represented in that figure. A change in the discount rate used in the utilitarian criterion leads to two changes in the figure; it alters the stationary value of s, s^*, which satisfies the condition $\delta u_1'(0) = u_2'(s^*)$, and in addition it alters the solution paths of the differential equations governing the movement of c and s, inter alia through changing the position of the curve on which $\frac{dc}{dt} = 0$. These equations, derived in chapter 3 and represented by the phase portrait in figure 6.3, are

$$u_1''(c_t)\dot{c}_t - \delta u_1'(c_t) = -u_2'(s_t)$$

$$\dot{s}_t = -c_t$$

So a reduction in the utilitarian discount rate increases the stationary stock of the resources, s^*, and also alters the phase portrait of the system by moving to the right the locus of points at which $\frac{dc}{dt} = 0$.

However, moving from the utilitarian solution to the Chichilnisky solution with the same discount rate in both achieves the first of these effects but not the second: it increases the stock of the resource conserved forever without altering in any way the phase portrait. It leads the optimum to follow one of the paths of the utilitarian phase portrait, but one with a different terminal condition from the utilitarian solution. The result is a path that is not a utilitarian optimum for any discount rate; the stationary stock would be optimal for a lower discount rate in the utilitarian context, but the dynamic path corresponds to the initial discount rate. The Chichilnisky approach introduces a degree of freedom not available in the utilitarian case; it breaks the link that exists in the utilitarian case between stationary solutions and dynamics. Formally,

PROPOSITION 19 Any path that is optimal for a utilitarian criterion is optimal for Chichilnisky's criterion for some value of the parameter α, but the converse is not true.

PROOF To see that any utilitarian optimum is a Chichilnisky optimum, just set $\alpha = 1$ in the Chichilnisky criterion. The argument above shows why the converse is not true.

6.4 Overtaking

How do these results compare with those from using the overtaking criterion, which, as we have seen, has some features in common with Chichilnisky's criterion? There is a very sharp difference: the overtaking criterion selects as optimal the green golden rule, the path along which nothing is consumed and the entire initial stock is maintained intact. This is easily established, using the fact that the overtaking criterion ranks paths with different limiting utility values according to those limiting values.

The argument is as follows. First, recall that the limiting utility value on any path is $u(0, s')$, where s' is the limiting resource stock. Provided that zero consumption does not incur an infinite utility penalty, this value is maximized by having the initial resource stock as the limiting stock. The program that consumes nothing has the largest limiting utility value and so overtakes any other program. The overtaking optimum is also the Rawlsian optimum and the green golden rule. It is furthermore the only solution with nondecreasing utility levels. This is a remarkable coincidence of views. Note, however, that complete conservation of the initial stock is optimal, whatever the size of the initial stock. The point is that nothing in these arguments depends on the size of the initial stock. Intuitively, this does not seem reasonable: complete conservation would make sense if the initial stock were very small, but not if it were very large. Intuitively, one feels that whether to conserve should depend on the size of the initial stock. Both the discounted utilitarian and the Chichilnisky optima have this property. This result is of course an illustration of the point made in chapter 5 that the overtaking criterion is more future-oriented than Chichilnisky's, and is close to being a dictatorship of the future. In fact, restricted to the family of paths that have limiting utility values, the overtaking criterion *is* a dictatorship of the future.

We can make one more useful observation on the overtaking criterion in this context. A stationary solution for the discounted utilitarian and Chichilnisky criteria is characterized by equality of the incremental costs and benefits from reducing consumption. This was the interpretation given to the equations (3.4) and (6.4): recall that in the utilitarian case characterized by (3.4), the interpretation is that the loss of utility from reducing consumption is the derivative of utility with respect to consumption times the drop in consumption, $u_1'(0)\Delta c$. The gain from an increased stock, which continues indefinitely, is the present value of the stream of incremental utilities accruing from an increased stock, $\int_0^\infty u_2'(s^*)\Delta c \exp(-\delta t)\, dt = \Delta c u_2'(s^*)/\delta$. Equality of the incremental gains and losses implies (3.4). Now, if the discount rate $\delta = 0$, as with the overtaking criterion, the gains to postponing consumption are always infinite. So one cannot characterize optimality by reference to equality of incremental costs and benefits from small changes in the path; this is a consequence of the fact that the overtaking criterion cannot be represented by a real-valued function.

Chapter 7
Renewable Resources Revisited

The next step in our analysis is to understand the implications of the various more future-oriented criteria such as Chichilnisky's in the case of renewable resources. The picture that emerges is rather different from the case of exhaustible resources: the renewability of the natural resource makes a fundamental difference to the structure of the problem. In the case of Chichilnisky's criterion, an optimal path exists only if the discount rate is nonconstant and falls to zero; if this is the case, then the optimal path asymptotes toward the green golden rule. An overtaking optimal path, if it exists, will also asymptote to the green golden rule, but in this case we cannot prove the existence of an optimal path. This is a consequence of the fact, noted in chapter 6, that the overtaking criterion cannot be represented by a real-valued function, so that the simplest proof of existence of optimal paths—to show that a continuous maximand is being maximized over a compact set (see Chichilnisky [19] and the appendix)—cannot be used. In the case of renewable resources, the green golden rule plays an important strategic role in the analysis: in addition to being a solution concept in its own right and the asymptote of two solution concepts (overtaking and Chichilnisky's), it has an interesting turnpike property, namely that finite horizon optima with zero discount rates and sufficiently long horizons spend a part of their duration arbitrarily close to the green golden rule.

We begin by studying the implications of Chichilnisky's criterion in the case of a constant rate of discount, and show that in this case, rather surprisingly, there is no optimum according to this criterion.

7.1 Constant Discount Rates

How does Chichilnisky's criterion alter matters when applied to an analysis of the optimal management of renewable resources? The formulation

of the dynamic constraints here is identical to that in chapter 4. With
the new objective, the problem now is to pick paths of consumption and
resource accumulation over time to

$$\max \alpha \int_0^\infty u(c_t, s_t)e^{-\delta t}\, dt + (1 - \alpha) \lim_{t \to \infty} u(c_t, s_t) \tag{7.1}$$

$$\text{s.t.} \quad \dot{s}_t = r(s_t) - c_t, \quad s_0 \text{ given}$$

For this framework, unlike the previous one, the change resulting from the
change in the criterion of optimality is substantial. With the Chichilnisky
criterion and the measure $\Delta(t)$ given by an exponential discount factor,
$\Delta(t) = e^{-\delta t}$, there is no solution to the overall optimization problem.
There is a solution only if $\Delta(t)$ takes a different, nonexponential form,
implying a nonconstant discount rate that tends asymptotically to zero.
Here Chichilnisky's criterion links in an unexpected way with the discussions of chapter 5 of individual attitudes toward the future; recall that we
noted empirical evidence that individuals making intertemporal choices
act as if they have nonconstant discount rates that decline over time.

Formally,

PROPOSITION 20 The problem of finding a path of use of a renewable
resource that is optimal according to the Chichilnisky criterion, that is,
problem (7.1), has no solution.

PROOF Consider first the utilitarian problem

$$\max \int_0^\infty u(c_t, s_t)e^{-\delta t}\, dt \quad \text{s.t.} \quad \dot{s}_t = r(s_t) - c_t, \quad s_0 \text{ given}$$

which is identical to that considered in chapter 4. The dynamics of
the solution is shown in figure 4.1, reproduced here as figure 7.1. The
utilitarian problem differs from the problem under consideration by the
lack of the term in limiting utility in the maximand.

Suppose now that the initial stock in (7.1) is s_0 in figure 7.1. By
proposition 17 of chapter 6, the solution to (7.1) must satisfy the first-
order conditions for optimality in the utilitarian problem. The solution
to the utilitarian optimality problem is to pick an initial consumption
level such that the initial stock and consumption levels together form
a point lying on the stable manifold of the stationary solution to the

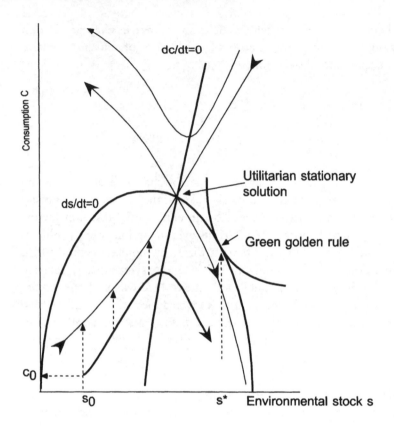

FIGURE 7.1 The dynamics of the utilitarian optimal use of a renewable resources.

utilitarian problem. Consider instead picking an initial value of c, say c_0, below the path leading to the saddlepoint, and follow the path from c_0 satisfying the utilitarian necessary conditions given in section 4.2 and repeated here:

$$u_1''(c)\dot{c} = u_1'(c)[\delta - r'(s)] - u_2'(s)$$

$$\dot{s}_t = r(s_t) - c_t$$

Call this path $(\bar{c}_t, \bar{s}_t)(c_0, s_0)$. Follow this path until it leads to the resource stock corresponding the green golden rule, that is, until the date t' such that on the path $(\bar{c}_t, \bar{s}_t)(c_0, s_0)$, $\bar{s}_{t'} = s^*$, and then at $t = t'$ increase consumption to the level corresponding to the green golden rule, that is, set $c_t = r(s^*)$ for all $t \geq t'$. This is feasible because $c_t \leq r(s_t)$

along such a path. Such a path is shown in figure 7.1. Formally, this path is $(c_t, s_t) = (\bar{c}_t, \bar{s}_t)(c_0, s_0)\,\forall\, t \leq t'$, where t' is defined by $\bar{s}_{t'} = s^*$, and $c_t = r(s^*), s_t = s^*\,\forall\, t \geq t'$.

Any such path satisfies the necessary conditions for utilitarian optimality up to time t' and leads to the green golden rule in finite time. It therefore attains the maximum of the term $\lim_{t\to\infty} u(c_t, s_t)$ over feasible paths. However, the utility integral that constitutes the first part of the maximand can be improved by picking a slightly higher initial value c_0 for consumption, again following the first-order conditions for optimality and reaching the green golden rule slightly later than t'. This is because such a path approximates more closely the utilitarian optimum. This does not detract from the second term in the maximand. By this process it is possible to increase the integral term in the maximand without reducing the limiting term and thus to approximate the independent maximization of both terms in the maximand: the discounted utilitarian term, by staying long enough close to the stable manifold leading to the utilitarian stationary solution, and the limit (purely finitely additive) term by moving to the green golden rule very far into the future.

Although it is possible to *approximate* the maximization of both terms in the maximand independently by postponing further and further the jump to the green golden rule, there is no feasible path that actually *achieves* the supremum of the sequence of payoff values. The supremum of the values of the maximand over feasible paths is approximated arbitrarily closely by paths that reach the green golden rule at later and later dates, but the limit of these paths never reaches the green golden rule and so does not achieve the supremum of the maximand. More formally, consider the limit as c_0 approaches the stable manifold of the utilitarian optimal solution of paths, that is, the limit as c_0 approaches the stable manifold of the paths

$$(c_t, s_t) = (\bar{c}_t, \bar{s}_t)(c_0, s_0)\,\forall\, t \leq t' \qquad \text{where } t' \text{ is defined by}$$
$$\bar{s}_{t'} = s^* \quad \text{and} \quad c_t = r(s^*), \qquad s_t = s^*\,\forall\, t > t'$$

On this limiting path $s_t < s^*\,\forall\, t$, so that the maximum of the limiting term is not attained. Yet the value of the integral term converges to its supremum along this sequence. Hence there is no solution to (7.1).

Intuitively, the nonexistence problem arises here because it is always possible to postpone further into the future moving to the green golden rule, with no cost in terms of limiting utility values but with a gain in terms of the integral of utilities. This is possible here, but not in the previous case, because of the renewability of the resource. But it is clear that a path that behaves like this cannot be optimal because the jump in the level of consumption means that the utility of an increment of consumption just before the jump exceeds that just after by a positive amount, so that total utility can be increased by shifting consumption from after to before the jump; this is true as long as the path is discontinuous, and so the utility integral along any such path can always be improved. The present-oriented part of Chichilnisky's criterion leads us to seek to postpone worrying about the long run into the long run, and in the limit we never worry about the long run. We have a solution to this problem only if we can resolve the conflict that apparently arises between the two parts of the criterion in the renewable resource case. This requires that the discount rate used in the integral part fall to zero in the long run, as we see in the next section.

7.2 Declining Discount Rates

With the Chichilnisky criterion formulated as

$$\alpha \int_0^\infty u(c_t, s_t) e^{-\delta t}\, dt + (1 - \alpha) \lim_{t \to \infty} u(c_t, s_t)$$

there is no solution to the problem optimal management of a renewable resource. In fact, as noted in chapter 5, the discount factor does not have to be an exponential function of time. The criterion can be stated slightly differently, in a way that is still consistent with Chichilnisky's axioms and is also consistent with solving the renewable resource problem. This reformulation builds on a point we have noted before: for the discounted utilitarian case, as the discount rate goes to zero, the stationary solution goes to the green golden rule. We therefore consider a modified objective function

$$\alpha \int_0^\infty u(c_t, s_t)\, \Delta(t)\, dt + (1 - \alpha) \lim_{t \to \infty} u(c_t, s_t)$$

where $\Delta(t)$ is the discount factor at time t and $\int_0^\infty \Delta(t)\, dt$ is finite. The discount rate $q(t)$ at time t implied by this formulation is minus the proportional rate of change of the discount factor:[1]

---

[1] A minus sign is needed because it is conventional to report the discount rate as a positive number, and the proportional rate of change of the discount factor Δ is negative.

$$q(t) = -\frac{\dot{\Delta}(t)}{\Delta(t)}$$

We assume that the discount rate goes to zero with t in the limit:

$$\lim_{t \to \infty} q(t) = 0 \qquad (7.2)$$

So the overall problem is now

$$\max \alpha \int_0^\infty u(c_t, s_t) \Delta(t) \, dt + (1 - \alpha) \lim_{t \to \infty} u(c_t, s_t)$$

$$\text{s.t.} \quad \dot{s}_t = r(s_t) - c_t, \qquad s_0 \text{ given}$$

where the discount factor $\Delta(t)$ is integrable and satisfies the condition (7.2) that the discount rate goes to zero in the limit. I will show that for this problem, there is a solution;[2] in fact, it is the solution to the utilitarian problem of maximizing just the first term in the above maximand, $\int_0^\infty u(c_t, s_t) \Delta(t) \, dt$. As before, we take the utility function to be separable in its arguments: $u(c, s) = u_1(c) + u_2(s)$. Formally,

PROPOSITION 21 Consider the problem

$$\max \alpha \int_0^\infty [u_1(c) + u_2(s)] \Delta(t) \, dt + (1 - \alpha) \lim_{t \to \infty} [u_1(c) + u_2(s)],$$

$$0 < \alpha < 1$$

$$\text{s.t.} \quad \dot{s}_t = r(s_t) - c_t, \qquad s_0 \text{ given}$$

where $\int_0^\infty \Delta(t) \, dt < \infty$, $q(t) = -[\bar{\Delta}(t)/\Delta(t)]$ and $\lim_{t \to \infty} q(t) = 0$. This problem has a solution that is identical to that which maximizes $\int_0^\infty [u_1(c) + u_2(s)] \Delta(t) \, dt$ subject to the same constraint. In other words, solving the utilitarian problem with the variable discount rate that goes to zero solves the overall problem.

PROOF Consider first the problem:

$$\max \alpha \int_0^\infty [u_1(c) + u_2(s)] \Delta(t) \, dt \quad \text{s.t.} \quad \dot{s}_t = r(s_t) - c_t, \qquad s_0 \text{ given}$$

We shall show that any solution to this problem approaches and attains the green golden rule asymptotically, which is the configuration of the

---

[2] I am grateful to Harl Ryder for suggesting this result and outlining the intuition behind it.

economy that gives the maximum of the term $(1 - \alpha) \lim_{t \to \infty} u(c_t, s_t)$. Hence, such a solution solves the overall problem.

The Hamiltonian for the integral problem is now

$$H = [u_1(c) + u_2(s)] \Delta(t) + \lambda_t \Delta(t)[r(s_t) - c_t]$$

and maximization of H with respect to consumption gives as before

$$u_1'(c_t) = \lambda_t$$

The rate of change of the shadow price λ_t is determined by

$$\frac{d}{dt}[\lambda_t \Delta(t)] = -[u_2'(s_t) \Delta(t) + \lambda_t \Delta(t)r'(s_t)]$$

The rate of change of the shadow price is therefore

$$\dot{\lambda}_t \Delta(t) + \lambda_t \dot{\Delta}(t) = -u_2'(s_t) \Delta(t) - \lambda_t \Delta(t)r'(s_t) \tag{7.3}$$

As $\dot{\Delta}(t)$ depends on time, this equation is not autonomous, that is, time appears explicitly as a variable. For such an equation, we cannot use the phase portraits and associated linearization techniques used before because the rates of change of c and s depend not only on the point in the c–s plane but also on the date. Rearranging and noting that $\dot{\Delta}(t)/\Delta(t) = -q(t)$, we have

$$\dot{\lambda}_t - \lambda_t q(t) = -u_2'(s_t) - u_1'(c_t)r'(s_t)$$

But in the limit $q = 0$, so in the limit this equation is autonomous; this equation and the stock growth equation form what is called in dynamic systems theory an asymptotically autonomous system (see Benaïm and Hirsch [15]). According to proposition 1.2 of [15], the asymptotic behavior and phase portrait of this nonautonomous system

$$\dot{\lambda}_t - \lambda_t q(t) = -u_2'(s_t) - u_1'(c_t)r'(s_t)$$
$$\dot{s}_t = r(s_t) - c_t \tag{7.4}$$

is the same as that of the autonomous system

$$\dot{\lambda}_t = -u_2'(s_t) - u_1'(c_t)r'(s_t)$$

$$\dot{s}_t = r(s_t) - c_t \tag{7.5}$$

which differs only in that the nonautonomous term $q(t)$ has been set equal to zero.[3] The pair of equations (7.5) is an autonomous system and the asymptotic stability properties of original system (7.4) are the same as those of the associated limiting autonomous system (7.5). This latter system can be analyzed by the standard techniques used before. At a stationary solution of (7.5), $\dot{\lambda}_t = 0$ and $c_t = r(s_t)$, so that

$$\frac{u_2'}{u_1'} = -r' \quad \text{and} \quad c_t = r(s_t)$$

which is just the definition of the green golden rule. Furthermore, by the arguments used in chapter 4, section 4.2, proposition 5, we can establish that the green golden rule is a saddlepoint of the system (7.5), as shown in figure 7.2.

So given the assumptions of the proposition, the optimal path for the problem

$$\max \int_0^\infty [u_1(c) + u_2(s)] \Delta(t)\, dt \quad \text{s.t.} \quad \dot{s}_t = r(s_t) - c_t, \qquad s_0 \text{ given}$$

is for any given initial stock s_0 to select an initial consumption level c_0 such that the trajectory from (c_0, s_0) approaches the green golden rule asymptotically.[4] But this path also leads to the maximum possible value of the term $\lim_{t \to \infty} [u_1(c) + u_2(s)]$, and therefore leads to a solution to the overall maximization problem.

Intuitively, one can see what drives this result. The nonexistence of an optimal path with a constant discount rate arose from a conflict between the long-run behavior of the path that maximizes the integral of discounted utilities and that of the path that maximizes the long-run utility level. When the discount rate goes to zero in the limit, that conflict is resolved. In fact, one can show that it is resolved only in this case, as stated by the following proposition.

---

[3]This equality is not always true; it requires locally uniform convergence of the nonautonomous system to the autonomous system. For details see Benaïm and Hirsch [15].

[4]Our arguments show this to be true in a neighborhood of the green golden rule; it is not clear to me that they establish a global result, although I strongly suspect that this is true.

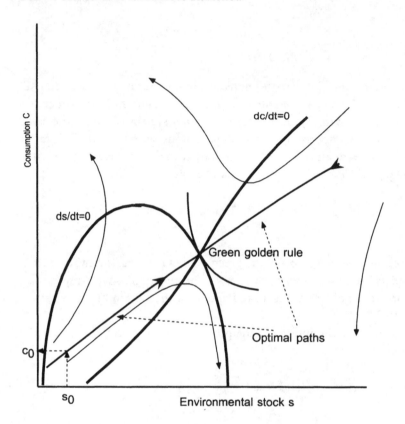

FIGURE 7.2 The green golden rule is the stationary solution of the autonomous system (7.5), and is a saddle point.

PROPOSITION 22 Consider the problem

$$\max \alpha \int_0^\infty [u_1(c) + u_2(s)] \Delta(t)\, dt + (1 - \alpha) \lim_{t \to \infty} [u_1(c) + u_2(s)], \quad 0 < \alpha < 1$$

$$\text{s.t.} \quad \dot{s}_t = r(s_t) - c_t, \qquad s_0 \text{ given}$$

where $\int_0^\infty \Delta(t)\, dt < \infty$ and $q(t) = -\dot{\Delta}(t)/\Delta(t)$. This problem has a solution if and only if $\lim_{t \to \infty} q(t) = 0$. In this case, the solution is identical to that which maximizes $\int_0^\infty [u_1(c) + u_2(s)] \Delta(t)\, dt$ subject to the same constraint.

PROOF The "if" part of this was proven in proposition 21. The "only if" part can be proven by an extension of the arguments in proposition 20, which established the nonexistence of solutions in the case of a constant discount rate. To apply the arguments there, assume contrary to the proposition that $\lim_{t\to\infty} \inf q(t) = \bar{q} > 0$, and then apply the arguments of proposition 18.

7.2.1 Examples — To complete this discussion, we review some examples of discount factors that satisfy the condition that the limiting discount rate goes to zero. The most obvious is

$$\Delta(t) = e^{-\delta(t)t}, \quad \text{with} \quad \lim_{t\to\infty} \delta(t) = 0$$

Another example[5] is

$$\Delta(t) = t^{-\alpha}, \quad \alpha > 1$$

Taking the starting date to be $t = 1$,[6] we have

$$\int_1^\infty t^{-\alpha}\, dt = \frac{1}{\alpha - 1}$$

and

$$\frac{\dot{\Delta}}{\Delta} = \frac{-\alpha}{t} \to 0 \quad \text{as} \quad t \to \infty$$

7.2.2 Empirical Evidence on Declining Discount Rates — Proposition 22 has substantial implications. It says that when we seek optimality with a criterion that is a complete ordering and is sensitive to the present and the long-run future, then with nonrenewable resources existence of a solution is tied to the limiting behavior of the discount rate; in the limit, we have to treat present and future utilities symmetrically in the evaluation of the integral of utilities. In a certain sense, the treatment of present and future in the integral has to be made consistent with the presence of the term $\lim_{t\to\infty}[u_1(c) + u_2(s)]$, which places positive weight on the very long run.

In chapter 5 we reviewed the growing body of empirical evidence that people actually behave like this in evaluating the future. We noted evidence that suggests that the discount rate people apply to future projects depends on, and declines with, the futurity of the project.

---

[5] Due to Harl Ryder.
[6] This discount factor is infinite when $t = 0$, hence the need to start from $t = 1$.

In chapter 5 we also noted that this empirically identified behavior is consistent with results from natural sciences that find that human responses to a change in a stimulus are nonlinear and are inversely proportional to the existing level of the stimulus. This is summarized in the Weber–Fechner law: *human response to a change in a stimulus is inversely proportional to the preexisting stimulus.* In symbols,

$$\frac{dr}{ds} = \frac{K}{s} \quad \text{or} \quad r = K \log s$$

where r is a response, s a stimulus, and K a constant. This has been found to apply to human responses to the intensity of both light and sound signals. We noted that the empirical results on discounting cited above suggest that something similar is happening in human responses to changes in the futurity of an event: a given change in futurity (e.g., postponement by one year) leads to a smaller response in terms of the decrease in weighting, the further the event already is in the future. In this case, the Weber–Fechner law can be applied to responses to distance in time, as well as to sound and light intensity, with the result that the discount rate is inversely proportional to distance into the future. Recalling that the discount factor is $\Delta(t)$ and the discount rate $q(t) = -\dot{\Delta}(t)/\Delta(t)$, we can formalize this as

$$q(t) = -\frac{1}{\Delta}\frac{d\Delta}{dt} = \frac{K}{t} \quad \text{or} \quad \Delta(t) = e^{K \log t} = t^K$$

Such a discount factor can meet all of the conditions we required above: the discount rate q goes to zero in the limit, the discount factor $\Delta(t)$ goes to zero, and the integral $\int_1^\infty \Delta(t)\,dt = \int_1^\infty e^{-K \log t}dt = \int_1^\infty t^{-K}\,dt$ converges for $K > 0$, as it always is. In fact, this interpretation gives rise to the second example of a nonconstant discount rate considered in the previous section. A discount factor $\Delta(t) = e^{-K \log t}$ has an interesting interpretation: the replacement of t by $\log t$ implies that we are measuring time differently, by equal proportional increments rather than by equal absolute increments.

7.3 Time Consistency

An issue raised by the previous propositions is that of *time consistency*. The concern here is the following. Consider a solution to an intertemporal optimization problem that is computed today and is to be carried out over some future period of time starting today. Suppose that the agent formulating the solution—an individual or a society—may at a future date recompute an optimal plan, using the same objective and the same constraints as initially but with initial conditions and starting date corresponding to those obtaining when the recomputation is done. We say that the initial solution is *time consistent* if this leads the agent to continue with the implementation of the initial solution. Another way of saying this is that a plan is time consistent if the passage of time alone gives no reason to change it.

We begin our analysis of time consistency with the simplest problem, the pure depletion problem as formulated by Hotelling and considered in chapter 2:

$$\max \int_0^\infty u(c_t)\, \Delta(t)\, dt \tag{7.6}$$

$$\text{with} \quad s_0 \text{ given,} \quad \dot{s}_t = -c_t, \quad s_t \geq 0 \,\forall t$$

Here $\Delta(t)$ is a general discount factor. Let $(c_t^*)_{0,\infty}$ be the time path of consumption that solves this problem. Consider also the same problem with a later starting date $T > 0$ and with an initial stock s_T given by the stock at date T on the path c_t^*:

$$\max \int_T^\infty u(c_t)\, \Delta(t)\, dt \tag{7.7}$$

$$\text{with} \quad s_T = s_0 - \int_0^T c_t^*\, dt, \quad \dot{s}_t = -c_t, \quad s_t \geq 0 \,\forall t$$

The second problem (7.7) is the problem that naturally arises if somewhere along the path chosen in response to the initial problem (7.6) we stop and recalculate our optimal path using as initial conditions those inherited at that date from the optimal path of problem (7.6). Let the solution to the second problem (7.7) be $(\hat{c}_t)_{T,\infty}$. Then:

DEFINITION 23 The solution to the problem (7.6) is time consistent if for all $t \geq T$, and for any $T, (c_t^*)_{T,\infty} = (\hat{c}_t)_{T,\infty}$.

This formalizes the idea that the passage of time alone gives no reason to revise the plan chosen initially. We return later to whether this is a desirable property; for the moment, our concern is to understand this property and to understand when solutions have it and when they do not.

Understanding time consistency requires that we think hard about the discount factor $\Delta(t)$, which shows how the weight on utilities at different dates changes over time. In particular, we have to answer the following question: if we are at date $t = 0$, then does the weight we place on date t, $\Delta(t)$, depend on the date t itself, or on some measure of the difference between t and the present? Using time zero as the present it is easy to confuse these approaches. But consider replanning at a date $t = T$, as time consistency requires. What is the discount factor to be applied to date $T_1 > T$? Is this discount factor a function of T_1, or of a measure of the difference or distance between T_1 and T? Obviously, it is the latter: we are discounting for distance in time, not for some concept of absolute time. Given this fact, what is the appropriate measure of the difference between T_1 and T?

The conventional measure is $(T_1 - T)$, but this is clearly not the only possibility: T_1/T seems to have a claim to validity also. These two numbers measure respectively the absolute and proportional differences between T_1 and T. Depending on which of these is selected, we have different insights into time consistency.

PROPOSITION 24 (1) When the discount factor applied to a future date is a function of the difference between that date and the present, then the solutions to problem (7.6) are time consistent if and only if the discount factor is of the form $\Delta(t) = e^{-\delta t}$ and in particular the discount factor applied to date T_1 at time T is $e^{-\delta(T_1-T)}$. (2) When the discount factor applied to a future date is a function of the ratio of that date to the present, then the solutions to problem (7.6) are time consistent if and only if the discount factor is of the form $\Delta(t) = e^{-\delta \log t}$ and in particular the discount factor applied to date T_1 at time T is $e^{-\delta \log T_1/T}$.

PROOF The proof of point (1) is standard and can be found in Heal [57]. The proof of point (2) is rather simple. The key point is that for time consistency the ratio of the weights applied to two dates, say, T_1 and T_2, should not depend on the date T at which the weights are calculated. But $e^{-\delta \log T_1/T}/e^{-\delta \log T_2/T} = e^{\delta \log T_2/T_1}$, which is independent of the date T. No other functional form for the dependence on time T gives this property.

So if we discount with absolute distance into the future, we need conventional exponential discounting to be sure of time consistent solutions.

If we discount with relative distance—as empirical evidence seems to suggest—then we need logarithmic discounting. If these conditions are not met, we have time inconsistent solutions.[7]

7.3.1 *Time Consistency and Chichilnisky's Criterion* — Our understanding of time consistency can now readily be applied to paths that are optimal according to Chichilnisky's criterion. We saw above that with renewable resources an optimal path exists if and only if the discount rate in the integral part of the maximand goes to zero in the limit, and in proposition 22 that a path is optimal according to this criterion if and only if it is optimal according to the utilitarian part of the maximand. From the preceding discussion of time consistency it immediately follows that with Chichilnisky's criterion and renewable resources we have a solution that is time consistent if and only if we discount as a function of relative rather than absolute distance into the future, and discounting is logarithmic with the discount factor given by $\Delta(t) = e^{-\delta \log t}$. Formally,

PROPOSITION 25 The solution to the problem of optimal management of a renewable resource with a discount rate falling asymptotically to zero:

$$\max \alpha \int_0^\infty [u_1(c) + u_2(s)] \Delta(t) \, dt + (1 - \alpha) \lim_{t \to \infty} [u_1(c) + u_2(s)], \quad 0 < \alpha < 1$$

$$\text{s.t.} \quad \dot{s}_t = r(s_t) - c_t, \quad s_0 \text{ given}, \quad \int_0^\infty \Delta(t) \, dt < \infty$$

$$q(t) = -\frac{\dot{\Delta}(t)}{\Delta(t)}, \quad \lim_{t \to \infty} q(t) = 0$$

is time consistent if and only if the discount factor applied to a future date is a function of the ratio of that date to the present, and is given by $\Delta(t) = e^{-\delta \log t}$.

PROOF This follows immediately from an application of the arguments in proposition 24.

7.3.2 *Asymptotic Time Consistency* — Although the paths that are optimal with respect to Chichilnisky's criterion in the case of renewable resources are not time consistent for all forms of intertemporal preferences, when they are inconsistent, their inconsistency is of a limited

---

[7]For an extensive discussion of time consistency and inconsistency in resource use problems, see Karp and Newberry [68]. They are interested in these issues in the context of multiagent problems and trying to model dynamic imperfectly competitive equilibria. See also Asheim [7].

and specific type. There are two distinctive characteristics. One is that whenever the path is recomputed, the optimal path still has as its positive limit set or asymptote the green golden rule: recomputation never changes the economy's long-run goals, about which there is unanimity among generations. Furthermore, the paths that solve (7.1) are asymptotically consistent, that is, the degree of time inconsistency decreases to zero along these paths. To see this, we need some notation. Let $_{T_1}\text{MRS}(c)_{T_2,T_3}$ be the marginal rate of substitution between consumption at dates T_2 and T_3 evaluated at T_1 on a path that is optimal for (7.1). Assume that $T_1 < T_2 < T_3$, and that we discount with differences rather than ratios in dates, as in (7.1), so that the solution is time inconsistent. Clearly

$$_{T_1}\text{MRS}(c)_{T_2,T_3} = \frac{\Delta(T_3 - T_1)u_1'(c_{T_3}^*)}{\Delta(T_2 - T_1)u_1'(c_{T_2}^*)}$$

We may define the marginal rate of substitution between stocks at dates T_2 and T_3 evaluated at T_1 correspondingly, denoting this $_{T_1}\text{MRS}(s)_{T_2,T_3}$. Now, a path is time consistent if and only if $_{T_1}\text{MRS}(c)_{T_2,T_3}$ and $_{T_1}\text{MRS}(s)_{T_2,T_3}$ are constant; we shall say that it is asymptotically consistent if the rates of changes of $_{T_1}\text{MRS}(c)_{T_2,T_3}$ and $_{T_1}\text{MRS}(s)_{T_2,T_3}$ go to zero. Formally,

DEFINITION 26 The path $\{c_t^*, s_t^*\}$ is asymptotically time consistent if

$$\lim_{T_1 \to \infty} \lim_{(T_2-T_1) \to \infty} \lim_{(T_3-T_1) \to \infty} \frac{d}{dT_1}[_{T_1}\text{MRS}(c)_{T_2,T_3}] = 0 \quad \text{and}$$

$$\lim_{T_1 \to \infty} \lim_{(T_2-T_1) \to \infty} \lim_{(T_3-T_1) \to \infty} \frac{d}{dT_1}[_{T_1}\text{MRS}(s)_{T_2,T_3}] = 0$$

Equipped with this definition, it is now simple to show that solutions to the Chichilnisky problem with renewable resources are asymptotically time consistent.

PROPOSITION 27 Consider the problem

$$\max \alpha \int_0^\infty u(c_t, s_t)\,\Delta(t)\,dt + (1 - \alpha) \lim_{t \to \infty} u(c_t, s_t)$$

$$\text{s.t.} \quad \dot{s}_t = r(s_t) - c_t, \qquad s_0 \text{ given}$$

where the discount factor $\Delta(t)$ is integrable and satisfies the condition (7.2) that the discount rate goes to zero in the limit. Any solution to this problem is asymptotically time consistent.

PROOF Recall that

$$\lim_{t \to \infty} \frac{1}{\Delta} \frac{d\Delta(t)}{dt} = 0.$$

It follows that as T_1, $T_2 - T_1$, and $T_3 - T_1$ tend to infinity, $(d/dT_1)_{T_1}\mathrm{MRS}(c)_{T_2,T_3}$ goes to zero.

7.3.3 Is Consistency Desirable? — Traditionally, welfare economists have always regarded time consistency as a very desirable property for intertemporal choices, both for individuals and for societies. More recently, this presumption has been questioned for individuals: philosophers and psychologists have noted that the same person at different stages of her or his life can reasonably be thought of as different people with different perspectives on life and different experiences.[8] On this interpretation, it is not unreasonable that one's trade-off between consumption at two different dates should depend on the date from which it is evaluated.

In this book, however, we are concerned with choices by societies rather than by individuals. Is time consistency a desideratum in this context? Probably the best way to get a perspective on this is to think of the selection of a plan for the use of environmental assets as an exercise in social choice. This is the perspective underlying Chichilnisky's approach to intertemporal equity: she uses an axiomatization drawn from approaches to social choice [20] (see also [49]). The parties to this choice are generations: the current generation and future generations. As time passes, today's present generation becomes a past generation. Identify each generation with a time period and index each by a natural number. The set of generations involved at time $T > 0$ is a subset of those that were involved at time 0, so that the parties to the social choice problem at time T are a subset of those who were parties to the problem at time 0. To be more precise, at time 0 we have a social choice problem to which the parties are $\{0, 1, 2, 3, \ldots, T, T + 1, \ldots\}$ whereas at time T the parties to the problem are $\{T, T + 1, T + 2, \ldots\}$. Time consistency requires that the choice in the second situation be the same as that in the first. In other words, it

---

[8]For a further discussion, see Harvey [56] and references therein, and also the volume edited by Lowenstein and Elster [78].

requires that as generations drop out of the choice process and move into the past, the outcome of this process remains unchanged. This is an unnatural requirement; it has no analogy in the social choice literature. There is no suggestion that a subset of a population should make the same choice as the whole population.

So from a social choice perspective, time consistency is a most unnatural requirement. Indeed, it is a condition that is very unlikely to be satisfied. Any intuitive appeal to the requirement of time consistency arises from a very strong view of what constitutes social rationality. It arises from a feeling that a rational society should make today choices about the future to which it will adhere in the future. This would undoubtedly simplify our lives, but we know from social choice theory and the general theory of preference aggregation that societies usually satisfy weaker rationality conditions than do the individuals of which they are composed. From this perspective, requiring time consistency is a move in the wrong direction: it is requiring that the aggregate satisfy a stronger condition than its individual constituents.

7.4 Overtaking

What are the implications of the overtaking criterion in the case of renewable resources? It is possible to describe what an optimal path would be like in this case, if it exists; however, it is not possible to be certain that it exists.[9] (In the case of the other criteria, existence either is obvious, as in the case of the green golden rule, or is proven in the appendix, as in the utilitarian and Chichilnisky cases.)

What would an overtaking optimal path look like, if it were to exist?

First note that by standard arguments, it must satisfy the utilitarian first-order conditions for optimality with the discount rate equal to zero:

$$u_1' = \lambda$$
$$\dot\lambda = -u_2' - u_1'r'$$
$$\dot s = r(s) - c$$

These conditions have as their saddlepoint a stable stationary solution the green golden rule.

Second, note that all paths satisfying the above first-order conditions have well-defined limiting utility values.

---

[9]Note that describing something does not prove that it exists; for example, I can describe a unicorn.

Finally, note that the green golden rule is the highest possible limiting utility value, so that the path satisfying the above conditions and approaching the green golden rule overtakes any other path satisfying the same first-order conditions.

Overall, then, it is clear that if an overtaking optimum exists, then it is a path that follows the utilitarian first-order conditions with a zero discount rate to the green golden rule. It has the same asymptote as a path optimal for Chichilnisky's criterion with a falling discount rate, and as it approaches this asymptote it has similar behavior. However, in its initial stretches it is more future oriented, which is consistent with what we saw in chapter 5 of the relationships between the two criteria.

7.5 Equal Treatment over Finite Horizons

Much of our ethical and moral intuition is grounded in the consideration of finite horizons. Life on earth will certainly be of finite duration, although it is difficult to determine its final date. It is therefore important to determine whether the phenomena we have been discussing are an artifact of infinite horizons, or have a clear relationship to approaches that seem reasonable in the context of finite horizons.

This section shows that these phenomena can be seen as an extension to infinite horizons of the properties of optimal solutions for an intuitively appealing criterion for finite horizons, the criterion that values all generations equally. This we call the finite equal treatment criterion. Indeed, for a general class of dynamic optimization problems, we will see that as the finite horizon increases, the optimal solutions of equal treatment finite horizon problems spend an increasing amount of time progressively closer to the green golden rule, which as we have seen is the asymptote for Chichilnisky-optimal and overtaking-optimal paths. In other words, the optimum according to Chichilnisky's criterion or according to the overtaking criterion determines a direction in which finite horizon equal treatment optima increasingly move as the horizon increases. We call this property a turnpike property.

DEFINITION 28 The equal treatment problem for horizon T is

$$\max \int_0^T u(c_t, s_t)\, dt \tag{7.8}$$

$$\text{s.t.} \quad \dot{s}_t = r(s_t) - c_t, \quad s_0 \text{ given}$$

Its solution is called the equal treatment optimum over T generations.

THEOREM 29 The green golden rule is the turnpike of finite horizon problems (7.8) in which each generation is treated equally. This means that as the number of generations T increases, the equal treatment optima for T generations spend some of their time increasingly close to a plan that is asymptotically approached by the optima for Chichilnisky's criterion (7.1) and by the optima according to the overtaking criterion. Formally, the distance[10] between the equal treatment optimal path for horizon T and the green golden rule g^* goes to zero as T goes to infinity.

PROOF Figure 7.3 illustrates the results. The solutions for the equal treatment criterion for finite time horizons T are indicated; these solutions are found by noting that the stock at T should be zero and then finding the appropriate initial conditions by solving the Euler–Lagrange equations. As $T \to \infty$, these paths comes closer and closer to the green golden rule. This is the turnpike property.

7.6 Summary

In the case of exhaustible resources, Chichilnisky's criterion leads to a solution that is in a certain sense between the utilitarian solution and the green golden rule: it naturally bridges a gap between the two solutions. It also enriches the set of possible solutions by decoupling the stationary solution reached from the dynamic path by which it is approached. In the case of renewable resources, it forces the two solutions (the utilitarian and the green golden rule) together and gives a solution only when these two solutions agree. This highlights the possibility, introduced by the renewability of the resource, of postponing further and further into the future any actions addressed primarily to the maximization of the long-run term $\lim u(c, s)$ in the objective function. This boosts the present welfare term $\int_0^\infty u(c_t, s_t)e^{-\delta t} \, dt$, and in the limit leads to the neglect of the long run altogether. Only when the two terms coalesce naturally, because the integral terms becomes more far sighted and future oriented with the passage of time because of the decline in the discount rate, can this conflict be resolved.

---

[10] As a measure of the distance between the path and point g^* I use the Hausdorf distance, the shortest distance between a point on the path and point g^*. Denoting the Hausdorf distance between points x and y by $HD(x, y)$, and the optimal path for horizon T by $\{Opt.T\}$, we have $HD(\{Opt.T\}, g^*) = \min_{x \in \{Opt.T\}} \|x, g^*\|$ where $\|x, y\|$ is the Euclidean distance between x and y.

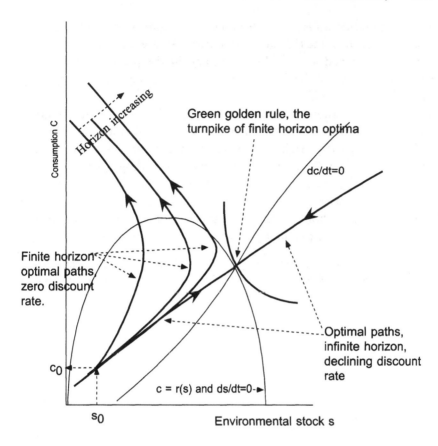

Consumption C

Horizon increasing

Green golden rule, the
turnpike of finite horizon optima

dc/dt=0

Finite horizon
optimal paths,
zero discount
rate.

Optimal paths,
infinite horizon,
declining discount
rate

c_0

c = r(s) and ds/dt=0-►

s_0

Environmental stock s

FIGURE 7.3 The green golden rule is the stationary state of an optimal path according to Chichilnisky's criterion and also the turnpike of finite horizon equal treatment (zero discount) optima.

For both exhaustible and renewable resources, maximization of the limiting term is attained by reaching the green golden rule;[11] in the case of exhaustible resources, this requires conservation of the entire stock, so that a choice between present and future has to be made right at the start of the planning period. Any act of consumption reduces the limiting utility level. In the renewable case, this is not true: there is not the same sharp conflict between present consumption and future welfare levels. Ironically, it is the absence of this conflict between present and future

---

[11]With exhaustible resources, use of the overtaking criterion leads to the green golden rule. It has not been possible to characterize the path that is optimal in the overtaking sense with renewable resources. It is natural to consider as a candidate the limit of the discounted utilitarian paths as the discount rate goes to zero. However, I am not aware of a proof that this limit exists or that if it existed it would be overtaking optimal.

consumption that leads to the phenomenon of nonexistence except in the case of a discount rate that declines to zero. We are tempted to optimize both the short run and the long run, and in the process run into the paradox described in this chapter: we neglect the long run by postponing forever attending to it.

A clear conclusion of the analysis of this chapter is that the green golden rule is an important configuration of the economy. It is clearly of interest in its own right, as giving the maximum sustainable utility level; it is also the asymptote of paths optimal according to Chichilnisky's criterion and the overtaking criterion (if such optima exist, which is not proven in the latter case). Furthermore, it has an interesting turnpike property for finite horizon equal treatment optima.[12] By aiming for the green golden rule, one cannot go far wrong in the long run.

---

[12]The zero discount rate utilitarian path that approaches the green golden rule is probably an "agreeable plan," in the sense of Hammond and Mirrlees [52]. See also Heal [57].

Chapter 8
Investment in a Backstop

We next study a problem different from either of those studied before: it has similarities to the depletion problem, but brings out more clearly the role of the value placed on the long run. We consider a society with a finite stock of a resource. This resource is nonrenewable, and once it is used up, there is no chance of finding more. As in the Hotelling problem, the resource is valuable only in consumption; the stock is not in itself a source of utility. A renewable replacement for this resource is available, but to produce it large fixed investments must be carried out now, cutting into current and future consumption. These investments will produce a flow of a renewable resource that will continue forever and is a complete substitute for the exhaustible resource, but will become available only in the far future. The question we pose is, How much should be invested now in the development of this alternative? What is the correct trade-off between present consumption and investment now to ensure the future availability of the backstop?

One can think of this as a model of decision making with respect to investment in the development of a replacement for fossil fuels, such as solar energy or geothermal power. In these cases, an investment now may lead to a renewable substitute for fossil fuels, or at least to a substitute in extremely abundant supply. But that substitute may be available only very far into the future, so that the payoffs to the investment are distant in time, even though costs must be incurred now. From a discounted utilitarian perspective, the costs today may dominate the equation; most of the benefits, being distant in time, are discounted heavily. Hence, the incentive to invest in the alternative energy source is small from the classical utilitarian point of view.

Another problem with this structure is investing in the secure disposal of nuclear waste. [1] The issue here is that some waste from nuclear power

---

[1] For a detailed discussion of this problem from the perspective of intertemporal optimality, see Wagner [108].

stations may remain dangerously radioactive for several centuries. It is very hard to devise disposal methods that will ensure that this waste is kept secure and will cause no contamination as long as it remains radioactive. However, it is possible to dispose of nuclear waste in ways that will keep it secure for about one century, but in most countries such options have been rejected on the grounds that they do not offer sufficient safeguards to the distant future. Obviously, such concerns seems to run counter to the discounted utilitarian paradigm and provide a justification for a main theme of this book, the investigation of alternative, more future-oriented frameworks. In the case of disposal of nuclear waste, ensuring no threat to generations far distant in the future requires a substantial investment now, with a payoff very far in the future. The payoff in this case consists of the avoidance of damage a century or more ahead.

Investing now to prevent global warming also has such a structure; replacing fossil fuels now and over the next half century by nonfossil energy sources represents an investment we make now to avoid a stream of damage that will begin only in the distant future. In this case, the time scale is shorter than that for nuclear waste: climate change experts expect that the earliest date at which greenhouse gas emissions could have a significant impact on global climate is about half a century ahead. The need to invest now, rather than waiting until much of that half century has elapsed, arises from the fact that there are extremely long lags in the response of the climate system to changes in economic policy toward greenhouse gas emission; it has been calculated that a reduction of greenhouse gas emissions to their 1990 levels by the year 2000, a very ambitious target indeed, would lead to no discernible improvement in global climate over the continuation of business as usual until the year 2040. [2]

Within a framework where investment today leads to a continuing flow of benefits in the far future, we characterize fully the investment and consumption policies that are optimal under discounted utilitarianism, the maximization of sustainable utility, overtaking, and Chichilnisky's criterion. The investment and consumption policies that are optimal in this later case are intermediate between those optimal under the discounted utilitarian and the green golden rule, and the optimal consumption path satisfies the usual local optimality conditions, which in this case are the Hotelling rule of chapter 2.

---

[2] For details, see Chichilnisky et al. [24].

8.1   The Model

Today we have a fixed stock s_0 of an exhaustible resource. By investing an amount i now we generate at a future date T a continuing flow $b(i)$ of a perfect substitute for this resource. There are positive but diminishing returns to investment, so that $b(i)$ is increasing in i, strictly concave, and $b(0) = 0$. Once started, the flow will continue at the same rate indefinitely. As suggested, we can think of this flow as a renewable alternative to an initial stock of a fossil fuel. Instantaneous consumption is evaluated by a strictly concave utility function $u(c)$. We assume that u has a constant elasticity η of marginal utility with respect to consumption, $\eta = -(cu''/u') > 0$. The challenge is to determine the appropriate balance between investment in the future technology and consumption now and in the future until the new technology is available.

The constraints on an optimal path change at the date T at which the renewable resource becomes available. Before that date we have an exhaustible resource, exactly as in the Hotelling problem of chapter 2:

$$\frac{ds_t}{dt} = -c_t, \qquad s_t - i \geq 0 \,\forall\, t \leq T \tag{8.1}$$

Subsequently, the remaining stock is augmented by the flow $b(i)$, which offsets the depletion due to consumption:

$$\frac{ds_t}{dt} = b(i) - c_t, \qquad s_t \geq 0 \,\forall\, t \geq T \tag{8.2}$$

From time zero to T we are depleting a fixed stock $s_0 - i$, the initial stock minus the amount invested in the backstop technology. This is the import of constraint (8.1). At T a permanent flow at rate $b(i)$ starts, at a rate that depends on what we invest today, and there is a remaining stock $s_T \geq 0$ from the initial endowment. This is captured by constraint (8.2). This framework is close to that of optimal depletion with a backstop technology studied in Dasgupta and Heal [40], and below we use several of the results from that paper. In [40], the date at which a backstop technology (the renewable resource) arrives is random, and in various extensions of that paper [42,43] we allow investment in R&D to affect the stochastic process that governs the arrival of the new technology. However, the scale on which the renewable resource becomes available is not affected by the R&D investment, and only discounted utilitarian approaches are studied. Here a key question is the appropriate level of the initial investment in the long-run technology.

8.2 Utilitarian Investment

The overall problem to be solved in the discounted utilitarian case is

$$\max \int_0^\infty u(c_t)e^{-\delta t}\, dt \quad \text{s.t. (8.1) and (8.2)} \tag{8.3}$$

where the maximization is with respect to the value of consumption c_t at all dates t, and with respect to the initial investment i in the backstop technology.

8.2.1 Optimality After the Backstop — Because of the change in the constraint at date T, at which the backstop becomes available, it is convenient to break the overall infinite horizon maximization problem (8.3) into two distinct parts, one from T to infinity and one from zero to T. We solve the overall problem by backward recursion, solving first the part from T to infinity.

The problem from date T on is

$$\max \int_T^\infty u(c_t)e^{-\delta(t-T)}\, dt \quad \text{s.t.}$$

$$\frac{ds_t}{dt} = b(i) - c_t, \qquad s_t \geq 0 \,\forall\, t \geq T \tag{8.4}$$

where $b(i)$ and s_T are given. The maximization is with respect to c_t for all $t \geq T$.

Let $V_T[s_T, b(i)]$ be the maximum value of the maximand in (8.4). This is naturally an increasing function of the flow of the substitute, $b(i)$, and of the stock remaining at time T, s_T, which are taken as given in problem (8.4). With this definition, which assigns a present utility value to the state of the economy at the date T of the transition, the overall infinite horizon problem (8.3) at time zero can now be written

$$\max \int_0^T u(c_t)e^{-\delta t}\, dt + V_T[s_T, b(i)]e^{-\delta T} \tag{8.5}$$

$$\text{s.t.} \ \frac{ds_t}{dt} = -c_t, \qquad s_t - i \geq 0 \,\forall\, t \leq T \tag{8.6}$$

We no longer need to specify the constraint from T on, as its impact is captured in the second term in the maximand (8.5), as in Dasgupta and Heal [40]. In problem (8.5) the maximization is with respect to the choice of i at date zero and the choice of c_t, $0 \le t \le T$. The choices of c_t and i determine the values of s_T and $V_T[s_T, b(i)]$ via equation (8.7):

$$s_T = s_0 - \int_0^T c_t \, dt - i \tag{8.7}$$

8.2.2 Optimality When the Stock Is Exhausted — In the solution to problem (8.4) from T on, there are two phases. In the first phase, which can be shown to be of finite length, a positive stock remains ($s_t > 0$) and it is possible to consume in excess of the flow $b(i)$. Consequently consumption satisfies the first-order conditions necessary for dynamic optimality and the associated shadow price of course follows Hotelling's rule.

In the second phase, which continues forever after the end of the first, none of the initial stock remains ($s_t = 0$) so that the flow $b(i)$ is the only source of the resource and the constraint $s_T \ge 0$ binds. One can readily verify that for optimality consumption is set at the maximum level, so that we have $c_t = b(i)$ during this phase.

Let T_1 be the date at which the second phase begins; that is, T_1 is the first date at which $s_t = 0$. The welfare accumulated during this second phase discounted back to date T is easily computed as

$$V_2[b(i)] = \int_{T_1}^{\infty} u[b(i)]e^{-\delta(t-T)}$$

8.2.3 Optimality Before the Stock Is Exhausted — The solution to the first phase problem from T on (8.4), that is, from T to T_1, is analyzed in Dasgupta and Heal ([40], pp. 6 and 7 and in particular propositions 1, 2, and 3). Consumption must satisfy

$$u'(c_t) = \xi_t$$
$$\dot{\xi}_t / \xi_t = \delta \tag{8.8}$$

where ξ is a shadow price. Consumption is therefore given by the equation

$$c_t = c_T^+ \exp\left[-\frac{\delta}{\eta}(t - T)\right]$$

The initial consumption level after the occurrence of the backstop c_T^+ and the date of exhaustion of the stock T_1 are determined by the following two equations:

$$\int_T^{T_1} c_t \, dt = s_T + (T_1 - T)b(i)$$

and

$$c_{T_1} = c_T^+ \exp\left[-\frac{\delta}{\eta}(T_1 - T)\right] = b(i)$$

The first of these requires that the resource stock remaining at T be exhausted by T_1. The second requires that consumption be continuous at the date T_1 at which the stock is exhausted. Let $V_1[b(i), s_T]$ be the welfare accumulated along an optimal policy over the interval (T, T_1), discounted back to T. We can clearly express the value of the state of the economy at date T when the backstop becomes available as the sum of the values V_1 and V_2 that have just been defined:

$$V_T[s_T, b(i)] = V_1[b(i), s_T] + V_2[b(i)]$$

8.3 Solving the Problem Recursively

The overall optimization problem (8.5) now takes the form

$$\max \int_0^T u(c_t)e^{-\delta t} \, dt + V_1[b(i), s_T]e^{-\delta T} + V_2[b(i)]e^{-\delta T} \tag{8.9}$$

$$\text{s.t.} \quad \frac{ds_t}{dt} = -c_t, \qquad s_t - i \geq 0 \, \forall t \leq T \tag{8.10}$$

Over the interval $(0, T)$ the solution to this must satisfy the usual first-order optimality conditions, which imply that [cf. (2.6)]

$$\frac{\dot{c}}{c} = -\frac{\delta}{\eta} \tag{8.11}$$

Let $s^* = s_0 - i - s_T$, and note that $\int_0^T c_t \, dt = s^*$. We have from (8.11) that

$$c_0 \int_0^T \exp\left[\left(-\frac{\delta}{\eta}\right)t\right] dt = s^*$$

that is,

$$c_t = \frac{s^* \frac{\delta}{\eta}}{1 - \exp\left[(-\frac{\delta}{\eta})T\right]} \exp\left(-\frac{\delta}{\eta}t\right) \forall\, t \leq T \qquad (8.12)$$

Let the utility integral along an optimal path over the interval $(0, T)$ be $V_0(s^*)$.

PROPOSITION 30 The overall discounted utilitarian problem (8.3) can be expressed as maximization of the sum of three functions with respect to the two variables i and s_T as follows:

$$\max_{i, s_T}\left\{ V_0(s_0 - i - s_T) + V_1[b(i), s_T]e^{-\delta T} + V_2[b(i)]e^{-\delta T}\right\} \qquad (8.13)$$

At interior solutions the optimal choices of these variables satisfy

$$V_0' = \frac{\partial V_1}{\partial s_T}e^{-\delta T} \qquad (8.14)$$

$$V_0' = \frac{\partial V_1}{\partial b}b'e^{-\delta T} + V_2'b'e^{-\delta T} \qquad (8.15)$$

The derivative V_0' is just the marginal utility of consumption at date zero. Similarly, V_2' is proportional to the marginal utility of consumption after T_1, and the derivatives of V_1 can also be shown to depend on the marginal utility at time T. If the marginal utility of consumption is bounded, and the derivative of $b(i)$ is also bounded, then the right-hand sides of equations (8.14) and (8.15) go to zero as $T \to \infty$. Hence we have

PROPOSITION 31 If the marginal utility of consumption and the derivative of $b(i)$ are bounded, then for large enough T, the solution to the overall discounted utilitarian problem (8.13) will involve a corner solution in which $i = 0$; in this case $b = 0$ and the original overall problem (8.5) is in effect a single infinite horizon problem.

The key implication of this proposition is that, within the discounted utilitarian framework, if the return $b(i)$ to the investment is far enough into the future, that is, T is large enough, then no investment will be made. Not surprisingly, this conclusion is at variance with those emerging from more future-oriented approaches.

More generally, consider the role of (8.15) in determining the optimal value of i. In figure 8.1 the horizontal axis shows the value of the initial investment i, which may be between zero and the entire initial stock s_0; the vertical axis shows the values of V_0' and $(\partial V_1/\partial b)b'e^{-\delta T} + V_2'b'e^{-\delta T}$, the payoffs to incremental consumption at time zero and to an increase in the investment respectively. As the initial investment i increases, initial consumption falls and so V_0' rises, and conversely with $(\partial V_1/\partial b)b'e^{-\delta T} + V_2'b'e^{-\delta T}$. The optimal value of the initial investment i is that at which the marginal valuations V_0' and $(\partial V_1/\partial b)b'e^{-\delta T} + V_2'b'e^{-\delta T}$ are equal. The figure also shows that as T increases, $(\partial V_1/\partial b)b'e^{-\delta T} + V_2'b'e^{-\delta T}$ falls and is eventually always below V_0'. In this case we have a corner solution with $i = 0$.

8.3.1 Investing for the Long Run: The Green Golden Rule — Suppose that in contrast to the discounted utilitarian approach, we focus solely

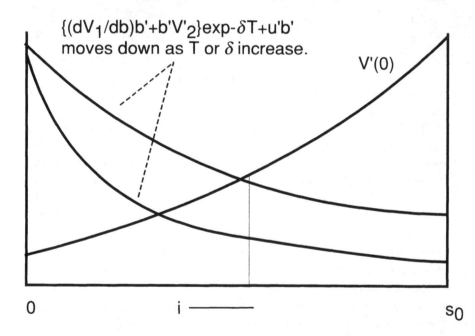

FIGURE 8.1 The optimal investment in the utilitarian case equates marginal utility at date zero with the marginal contribution to welfare at date T. The latter is discounted and falls as T increases. So for T sufficiently large there may be a corner solution with $i = 0$.

on the very long run and seek the green golden rule solution. Formally, the aim in this case is to

$$\max \lim_{t \to \infty} u(c_t) \tag{8.16}$$

The solution is simple: *invest the entire initial stock in the backstop*, as $\lim_{t \to \infty} u(c_t) = u[b(i)]$, so that we have to maximize $b(i)$ subject to the constraint $i \le s_0$. There will be no consumption before date T. Society abstains from consumption up to date T to take maximum advantage of the return from investment in the backstop, which will then continue forever. Maximizing any increasing function of the limiting utility value would of course lead to the same solution. In particular, the outcome would be the same if we maximized long-run average utility or the lim inf of utility values.

8.3.2 The Rawlsian Solution — In several previous cases, we have seen that the Rawlsian solution and the green golden rule coincide. What is their relationship in the present framework? Because investing everything in the backstop technology leaves nothing for consumption before its arrival, the worst-off generations are those before the backstop. Investing less in the backstop makes them better off; therefore, investing for the long run is not the same as choosing the Rawlsian optimum.

The Rawlsian optimum in the present model involves the maximum feasible constant consumption level, attained by choosing a level of investment in the backstop i_R that satisfies

$$b(i_R) = \frac{s_0 - i_R}{T}$$

With this investment level it is possible to maintain a constant consumption level over time: the constant rate $(s_0 - i_R)/T$ is feasible up to T, and $b(i_R)$ is feasible after T. Under the assumptions maintained here, there is a unique value i_R satisfying this condition and solving the Rawlsian problem. Such an outcome is inefficient with respect to any discounted integral criterion, although it does seem in the current context to capture some notions of sustainability and intergenerational equity.

8.3.3 Investing for Present and Future — Chichilnisky's definition of an optimal path is one that solves the problem

$$\max \alpha \int_T^\infty u(c_t)e^{-\delta(t-T)}\, dt + (1 - \alpha) \lim_{t \to \infty} u(c_t)$$

In the present case the limiting utility value is just $u[b(i)]$. Hence the maximand can be expressed as

$$\alpha V_0(s_0 - i - s_T) + \alpha V_1[b(i), s_T]e^{-\delta T} + \alpha V_2[b(i)]e^{-\delta T} + (1 - \alpha)u[b(i)] \tag{8.17}$$

In the previous results we characterized the investment and consumption levels that maximize each component of the maximand separately. Next we put the two together. This is done graphically in figure 8.2, a modification of figure 8.1. Figure 8.1 characterized the level of investment i that maximizes the integral of discounted utilities; this is the level at which $(\partial V_1/\partial b)b'e^{-\delta T} + V_2'b'e^{-\delta T}$ and V_0' are equal, or is a corner solution with $i = 0$ if $(\partial V_1/\partial b)b'e^{-\delta T} + V_2'b'e^{-\delta T}$ is uniformly below V_0'. Now we have to add to $\alpha(\partial V_1/\partial b)b'e^{-\delta T} + \alpha V_2'b'e^{-\delta T}$ the derivative of $(1 - \alpha)u[b(i)]$ with respect to i, that is, $(1 - \alpha)u'b'$. This term is *not* discounted, and so is unaffected by the value of T.

A maximum of (8.17) is characterized by equality of $\alpha V_0'$ with $\alpha(\partial V_1/\partial b)b'e^{-\delta T} + \alpha V_2'b'e^{-\delta T} + (1 - \alpha)u'b'$. This clearly involves a higher level of investment than that characterizing the maximization of the discounted sum of utilities, and a lower level than that which maximizes the limiting value of utility. Note that by the construction of

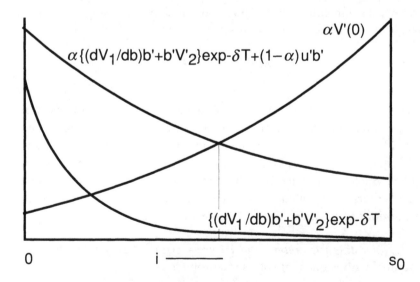

FIGURE 8.2 Adding a term depending on undiscounted limiting utility increases the optimal investment in the backstop technology.

the terms $V_0(s_0 - i - s_T)$ and $V_1[b(i), s_T]e^{-\delta T}$, the optimal path in this case satisfies the Hotelling rule or local optimality conditions until the first date at which $s_T = 0$, that is, until T_1. We have therefore established the following:

PROPOSITION 32 The optimal level of investment according to the Chichilnisky criterion is greater than that which maximizes the discounted sum of utilities, and less than that which maximizes the limiting utility value. Furthermore, the optimal consumption path satisfies the conditions necessary for a solution to the discounted utilitarian problem.

It is not clear whether this involves more or less investment than the Rawlsian criterion; however, the outcome is intertemporally efficient.

8.3.4 Overtaking — In the present framework, it is straightforward to see that the overtaking optimal path, as in the case of the pure depletion problems of chapter 6, is the same as the green golden rule. This path maximizes the limiting utility value and will overtake all other feasible paths.

8.3.5 Choosing When to Invest — An extension of the problem already solved is to allow the date at which the investment i is made to be chosen as part of the overall solution. All other aspects of the problem remain unaltered—the function $b(i)$ and the delay T between investment and the backstop—but the investment in the backstop is now i_t, where t is the date at which the investment is made. In this case it is straightforward to show the following:

- Under the discounted utilitarian regime, if an investment is made in the backstop, it will be made at time $t = 0$.
- Under the regime of maximizing limiting utility values, the timing of the investment is indeterminate.
- Under the regime of investing for present and future, the Chichilnisky criterion, the investment, if made, will again be made at time $t = 0$, and will be greater than the investment under the purely utilitarian regime.

The reasoning is simple. Postponing the investment reduces the utilitarian benefits by pushing them into the future and causing them to be more heavily discounted, but leaves the costs unchanged. The investment still must be financed out of the initial stock s_0 of the resource, and the opportunity cost of a unit of investment from that stock, at whatever date, is still V_0' in the notation of equation (8.14). The investment is therefore most attractive if carried out immediately.

8.4 Conclusions

For the problem studied in this chapter, the solution responds naturally and unsurprisingly to changes in the specification of the objective function: as we place more emphasis on the future, we see an increase in the initial investment in a technology that pays off only in the long run. In this respect the problem is like the exhaustible resource problem of chapters 3 and 6 rather than the renewable resource problem of chapters 4 and 7. It involves the creation of a renewable resource, but the key constraint about which decisions must be made arises from the exhaustibility of a resource. Present and future play a zero-sum game and are in direct competition. In such cases, it seems, one does not need to restrict the behavior of the discount rate to ensure the existence of a solution.

Part III
Capital Accumulation

Chapter 9
Exhaustibility and Accumulation

Previous chapters presented several possible alternative formalizations of sustainability in the context of economies with nonrenewable or renewable resources, but always without production or the accumulation of capital. This could in principle be a serious limitation to their relevance to a policy debate, but it was nonetheless essential in developing an intuition about the logical structure of propositions available in this field. Clearly the accumulation of produced capital can in some areas substitute for the availability of natural resources; for example, more capital, and more efficient capital can substitute for oil and gas, just as desalination plants can substitute for fresh water. But as noted in chapter 1, there are also many areas in which replacing natural capital by produced capital does not leave us as well off as before. An analysis of the implications of substitutability between natural and produced capital was a major theme of Dasgupta and Heal [40,41]. There it was shown that for certain values of the elasticity of substitution between natural resources and produced capital, and certain behaviors of the limiting value of the marginal product of capital as the capital–resource ratio increases, this possibility can indeed alter fundamentally the growth possibilities open to the economy. Here the matter is more complex because we are concerned with a natural resource that is valued in its own right and may be renewable and so have its own dynamics.

Equipped with a good understanding of how different criteria and constraints interact in simplified contexts, it is time to address the same issues in the more demanding environment of an economy with production and capital accumulation. This is a much more complex endeavor: geometric representations are no longer available, because in studying the dynamics of optimal paths we are typically dealing with systems of four simultaneous differential equations. Despite this complexity, it is possible to establish results that are in an obvious sense extensions of the results of previous chapters. The principles we have established in the simpler models of earlier chapters carry through to these more general cases, although in general we cannot extract as much detail about solutions analytically as we can in the simpler cases. It is reassuring that the intuitions and the key qualitative features of the results for the earlier simplified cases carry over the more general frameworks. This shows that they are robust to some of the simplifying assumptions we have made. In this chapter we address these issues in the context of a model with capital accumulation and exhaustible resources; in chapter 10, we consider the same issues when the resource is renewable.

The model we shall use is an extension of that introduced by Dasgupta and Heal in 1974 [40]; that model embeds the Hotelling problem of chapter 2 in a version of the Solow growth model. A production function $F(k, \sigma)$ gives the output of a produced good, which can be invested in capital formation or used in consumption, as a function of the input of a capital stock k and the rate of consumption σ of a nonrenewable resource, which is used as an input to production. The stock of the resource is denoted by s. So the national income identity in this model is just

$$\dot{k}_t = F(k_t, \sigma_t) - c_t$$

where c_t is the consumption of output at time t and $\dot{k}_t$ the rate of capital accumulation at time t. As before, utility is given by a function $u(c_t, s_t)$ and as is now customary depends on a flow of consumption and on the remaining stock of the resource.

9.1 The Utilitarian Optimum

The utilitarian problem in this context is therefore simply

$$\max \int_0^\infty u(c_t, s_t) e^{-\delta t} \, dt \tag{9.1}$$

$$\dot{k}_t = F(k_t, \sigma_t) - c_t \quad \text{and} \quad \dot{s}_t = -\sigma_t, \qquad s_t \geq 0 \, \forall \, t$$

Dasgupta and Heal give a complete analysis of this problem when utility depends only on current consumption. Krautkraemer [71] analyses some aspects of the present case. A key distinction introduced in Dasgupta and Heal, which will prove to be critical in the current context also, is between situations in which the resource is necessary as an input and those in which it is not. We say that the resource input is *necessary* if $F(k_t, 0) = 0$, whatever the value of k_t. Otherwise it is *unnecessary*. So if a resource is necessary, there is no production without it. Intuitively it is obvious what is at stake here and why it matters.[1] If the resource is necessary, society must make a choice: generally, present and future play a zero-sum game and are in competition (but see footnote 1). This choice is less stark if the resource is renewable, a case we consider later.

The first-order conditions for a solution to (9.1) are obtained from the Hamiltonian

$$H = u(c, s)e^{-\delta t} + \lambda e^{-\delta t}[F(k, \sigma) - c] - \mu e^{-\delta t} \sigma$$

where λ and μ are respectively the shadow prices associated with the capital stock k and the resource stock s. Optimality requires that

$$u_c = \lambda \tag{9.2}$$

$$\lambda F_\sigma = \mu \tag{9.3}$$

$$\dot{\lambda} - \delta\lambda = -\lambda F_k \tag{9.4}$$

$$\dot{\mu} - \delta\mu = -u_s \tag{9.5}$$

All these conditions are quite intuitive: the shadow price of the produced good must equal its marginal utility in consumption (9.2), the shadow price of the resource must equal its marginal productivity in production times the shadow price of what it produces (9.3), the third condition (9.4) is Ramsey's rule requiring that the shadow price of the capital good rise at a rate given by the difference between the discount rate and the marginal product of capital ($\dot{\lambda}/\lambda = \delta - F_k$), and the final condition (9.5) reduces to the familiar Hotelling rule if the resource stock has no utility in consumption (i.e., $u_s = 0$). Hotelling's rule is familiar and needs no further comment. Ramsey's rule, dating back to Ramsey's 1928 paper [91], which provided the foundations of optimal growth theory, also has a straightforward intuition. It implies that in equilibrium the marginal product of capital F_k, which is the return to waiting, equals the discount rate δ, which is the cost of waiting.

---

[1]The Cobb–Douglas production function is a borderline case, as shown in Dasgupta and Heal [41], chapter 7. The resource is always necessary for production, but a positive level of production may continue forever.

9.1.1 Stationary Solutions — In this more complex context, we can ask the same questions as were asked, and answered, of the simpler models. The first is whether there is a stationary configuration of the economy that satisfies the conditions necessary for optimality according the utilitarian criterion. We assume that the marginal utility of consumption at zero is finite: $u_c(0, s) < \infty$ for any $s > 0$. This assumption is important in analyzing stationary solutions when the resource is essential to production. At a stationary solution, the rate of resource use must be zero, that is, $\sigma = 0$, if the stock s is to be constant. For the capital stock to be constant from (9.1) we need the level of consumption to equal output, that is, $c = F(k, 0)$. From equations (9.4) and (9.5) for the shadow prices we see that one also needs $\delta = F_k$ and $\delta\mu = u_s$ at a stationary solution. Using in addition the first order conditions (9.2) and (9.3), we see that at a stationary solution satisfying the optimality conditions

$$\frac{u_s[F(k, 0), s]}{u_c[F(k, 0), s]} = \delta F_\sigma(k, 0) \tag{9.6}$$

and

$$F_k(k, 0) = \delta \tag{9.7}$$

The first equation here (9.6) is familiar: it is a generalization of the equation that characterized a stationary solution in the pure depletion case of chapter 3, equation (3.4), which was just $\frac{u_s}{u_c} = \delta$, requiring equality between the marginal rate of substitution between stock and flow and the discount rate. With no production, the productivity of an increment of the resource stock in producing consumption is just unity, and these two equations are identical. Equation (9.6) can therefore be given the same kind of interpretation as equation (3.4) in terms of the equality of the incremental contributions to stock or flow. Rewrite it as

$$u_c F_\sigma = \frac{u_s}{\delta}$$

The return to consuming an incremental unit of the resource is $u_c F_\sigma$, its impact on production times the impact on utility of the output it produces. The return on adding to the stock is u_s/δ, which is the present discounted value of the stream of incremental utilities from an increase in the stock. At a stationary optimal configuration, the returns to incremental changes in consumption c and the stock s must be equal. Another interpretation of this equation is that the marginal rate of substitution between stock and flow is equated to the discount rate: here the marginal rate of substitution between stock and flow is $u_s/u_c F_\sigma$, as the conversion of a unit of stock into flow requires its use in production. Equation (9.7) is entirely standard from the neoclassical theory of optimal growth (for example, see Heal [57]); it requires that the marginal product of capital equal the discount rate.

We have characterized a stationary solution to the utilitarian optimality conditions; this leaves open the question of whether such a solution exists. Equations (9.6) and (9.7) form two equations in two unknowns, k and s. The second equation fixes k, say at k^*, and then the first fixes s at s^*. So in general there is a unique stationary solution to the utilitarian optimality conditions. It is characterized by equality of the marginal product of capital to the discount rate and equality of the marginal rate of substitution between stock and flow to the marginal rate of transformation between these in production—all very intuitive. If the resource is not necessary in production, then $F(k, 0) > 0$ and consumption is positive at the stationary solution, given by $c = F(k^*, 0)$, where k^* satisfies $F_k(k^*, 0) = \delta$; in the essential case $F(k, 0) = 0$ and consumption is zero. The effect of a decrease in the discount rate depends on which case we are in. With the resource not necessary, a drop in δ leads to a larger stationary capital stock and hence stationary consumption; it can also be shown under certain fairly general conditions on the production function and the utility function to lead to a higher stationary value of the resource stock. With the resource necessary, a drop in the discount rate still leads to a higher stationary capital stock, but not to a higher stationary consumption, which is zero in this case. Again, it generally leads to a higher stationary level of the resource stock. To summarize,

PROPOSITION 33 Consider an economy with an exhaustible resource as an input to production. Assume that the marginal utility of the produced good at zero consumption is finite. Then the utilitarian optimality conditions admit a stationary solution at which a positive amount of the resource is preserved forever. The amount of the resource preserved forever is determined by equations (9.6) and (9.7). Consumption at the stationary solution is positive if the resource is unnecessary for production and zero otherwise.

9.1.2 Dynamics of the Utilitarian Optimum — The dynamics of the system of equations describing the utilitarian solution can be studied in the neighborhood of a stationary solution by the classic method of linearizing the system at the stationary solution and then studying the eigenvalues and eigenvectors of the matrix of the linearized system. The relevant system consists of the evolution equations for capital and the resource from (9.1), and the evolution equations for the associated shadow prices (9.4) and (9.5), and is

$$\dot{k}_t = F[k_t, \sigma(\mu_t, \lambda_t, k_t)] - c(s_t, \lambda_t)$$

$$\dot{s}_t = -\sigma(\mu_t, \lambda_t, k_t), \qquad \text{with } s_t \geq 0 \, \forall \, t$$

$$\dot{\lambda}_t - \delta\lambda_t = -\lambda_t F_k[k_t, \sigma(\mu_t, \lambda_t, k_t)]$$

$$\dot{\mu}_t - \delta\mu_t = -u_s[c(s_t, \lambda_t), s_t]$$

Here variables c and σ are expressed as functions of k, s, λ, and μ, determined by first-order conditions (9.2) and (9.3). Treating (9.2) and (9.3) as implicit functional relations,

$$u_c(c, s) - \lambda = 0, \qquad \lambda F_\sigma(k, \sigma) - \mu = 0$$

we note that

$$\frac{\partial c}{\partial \lambda} = \frac{1}{u_{cc}}, \qquad \frac{\partial c}{\partial s} = -\frac{u_{cs}}{u_{cc}}$$

and

$$\frac{\partial \sigma}{\partial k} = -\frac{F_{\sigma k}}{F_{\sigma\sigma}}, \qquad \frac{\partial \sigma}{\partial \lambda} = -\frac{F_\sigma}{\lambda F_{\sigma\sigma}}, \qquad \frac{\partial \sigma}{\partial \mu} = \frac{1}{\lambda F_{\sigma\sigma}}$$

Using these derivatives, the linearized system evaluated at the stationary solution is

$$
\begin{bmatrix} \dot{k} \\ \\ \dot{s} \\ \\ \dot{\lambda} \\ \\ \dot{\mu} \end{bmatrix}
=
\begin{bmatrix}
\delta - F_\sigma \dfrac{F_{\sigma k}}{F_{\sigma\sigma}} & \dfrac{u_{cs}}{u_{cc}} & -\dfrac{F_\sigma F_\sigma}{\lambda F_{\sigma\sigma}} - \dfrac{1}{u_{cc}} & \dfrac{F_\sigma}{\lambda F_{\sigma\sigma}} \\ \\
\dfrac{F_{\sigma k}}{F_{\sigma\sigma}} & 0 & \dfrac{F_\sigma}{\lambda F_{\sigma\sigma}} & -\dfrac{1}{\lambda F_{\sigma\sigma}} \\ \\
-\lambda F_{kk} + \dfrac{\lambda F_{k\sigma} F_{\sigma k}}{F_{\sigma\sigma}} & 0 & \dfrac{F_{k\sigma} F_\sigma}{F_{\sigma\sigma}} & -\dfrac{F_{k\sigma}}{F_{\sigma\sigma}} \\ \\
0 & \dfrac{u_{sc} u_{cs}}{u_{cc}} - u_{ss} & -\dfrac{u_{sc}}{u_{cc}} & \delta
\end{bmatrix}
\begin{bmatrix} k \\ \\ s \\ \\ \lambda \\ \\ \mu \end{bmatrix}
$$

In general it is hard to determine the signs of the real parts of the roots of this matrix. To obtain sharp results on the eigenvalues and eigenvectors, we need to make some simplifying assumptions. One assumption that yields sharp results is that $F_{\sigma\sigma}$ is large, so that terms divided by this can be neglected. This amounts to assuming that the marginal productivity of the resource in production falls rapidly; it is important to have small quantities of the resource as an input, but using more does not contribute significantly to output. With this assumption the matrix simplifies to

$$
\begin{bmatrix}
\delta & \dfrac{u_{cs}}{u_{cc}} & -\dfrac{1}{u_{cc}} & 0 \\[2ex]
0 & 0 & 0 & 0 \\[2ex]
-\lambda F_{kk} & 0 & 0 & 0 \\[2ex]
0 & \dfrac{u_{sc}u_{cs}}{u_{cc}} - u_{ss} & -\dfrac{u_{sc}}{u_{cc}} & \delta
\end{bmatrix}
$$

The eigenvalues in this case are

$$
\delta, 0, \frac{\delta}{2} \pm \frac{1}{2u_{cc}}\sqrt{u_{cc}^2\delta^2 + 4\lambda F_{kk}u_{cc}}
$$

Two eigenvalues are positive, and in addition there is always one real negative eigenvalue (because $4\lambda F_{kk}u_{cc}$ is positive under conventional assumptions about the production and utility functions) and one zero eigenvalue to the matrix of the linearized system. The eigenvector associated with the eigenvalue zero is $(0, 1, 0, 0)$, the unit vector on the s axis. Because the rate of change of s has to be nonpositive, it is clear that a stationary solution cannot be stable or unstable in the usual sense in this dimension, so that there must be a zero eigenvalue associated with this dimension. The stationary solution of the system is nevertheless stable in this dimension, in that it can be approached only from above, and from this direction it is stable. Hence there is a two-dimensional submanifold from which the utilitarian stationary solution can be approached and is stable, given that s_t is nonincreasing and bounded below on feasible paths. In summary,

PROPOSITION 34 If $F_{\sigma\sigma}$ is large, so that the marginal productivity of the resource falls rapidly, the utilitarian stationary solution is locally a saddle point.

There are other cases in which the stationary solution is locally a saddle point. We leave those as an exercise to the reader.

9.2   The Green Golden Rule

What configuration of the economy gives the maximum sustainable utility level, which we earlier called the green golden rule? Consider first the case of a resource necessary for production. The only constant level of the resource input σ that is feasible at a stationary state is zero, in which case the only level of output maintainable at a stationary state is zero because the resource is necessary for production. Hence utility in the long run is given by $u(0, s)$ and so is maximized by maintaining the entire initial stock s_0 of the resource intact, as in the case of depletable resources in chapter 3. Formally:

PROPOSITION 35 With an exhaustible resource that is necessary to production, the maximum sustainable utility level is attained by conserving the entire stock and consuming none of the produced good.

The case of a resource that is not necessary to production is more complex. When the resource is not needed for production, $F(k, 0) > 0$ for $k > 0$, so that it is possible to build up the capital stock and the level of consumption without ever using the resource. We have to distinguish two cases.

Case 1: Output as a function of capital stock k is bounded when the resource input is zero, so that $\exists b > 0: F(k, 0) \leq b \forall k$. Let $b = \sup_k F(k, 0)$, and assume also that k^* is such that $F(k^*, 0) = b$.

Case 2: Output as a function of capital stock is unbounded when the resource input is zero: $\forall b \exists k(b), F[k(b), 0] > b$.

In case 1 we can look for the greatest sustainable utility level. This is given by $u(b, s_0)$ where s_0 is the initial stock of the resource. The green golden rule policy is to maintain the initial stock of the resource intact and accumulate capital to the level k^*, and then consume the entire output.

Case 2 has two subcases. In the first subcase utility is increasing without limit in consumption at a given level of the resource stock. This means that unbounded utility levels can be attained with no depletion of the resource. Such a possibility seems to contradict the problem from which the literature on sustainability derives. In this case there is no maximum sustainable utility level. In the second subcase utility is eventually declining in consumption at the initial resource stock, as shown in figure 9.1.

In such a case, we have a maximum sustainable utility level. So formally,

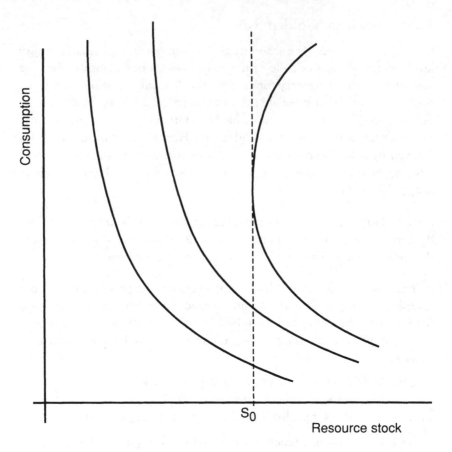

FIGURE 9.1 Preferences showing satiation with respect to consumption at the initial stock level. In this case the green golden rule is well defined even if the resource is not essential.

PROPOSITION 36 With an exhaustible resource that is not necessary to production, we must distinguish two cases. In case 1 (defined above) there is a maximum sustainable utility level that is attained by preserving the initial stock of the resource intact and accumulating capital up to the level k^*, and then consuming the entire output. In case 2, the maximum sustainable utility level is well defined only if utility is satiated in consumption at a finite consumption level c^* at the initial resource stock. In this case, maximum sustainable utility is obtained by preserving the entire stock and accumulating capital to the level k^* defined by $c^* = F(k^*, 0)$.

9.3 The Chichilnisky Criterion

The utilitarian criterion places little weight on the long run future; the green golden rule places weight only on the very long run, perhaps going too far in the opposite direction. As we have seen from previous chapters, Chichilnisky's criterion, developed in chapter 5, combines elements of both, and tends to combine them in the solutions it produces. In particular, in the problem of depletion of a stock without production that was considered in chapter 6, Chichilnisky's criterion led to an outcome that was in a clear sense intermediate between those produced by the green golden rule and the discounted utilitarian framework. Is this still true with the addition of capital accumulation and production?

According to Chichilnisky's criterion, the problem in hand is as follows:

$$\max \alpha \int_0^\infty u(c_t, s_t)e^{-\delta t} \, dt + (1 - \alpha) \lim_{t \to \infty} u(c_t, s_t) \tag{9.8}$$

$$\dot{k}_t = F(k_t, \sigma_t) - c_t \quad \text{and} \quad \dot{s}_t = -\sigma_t, \quad \text{with } s_t \geq 0 \, \forall t$$

As before, the criterion here is not amenable to optimization by the usual techniques of calculus of variations and their derivatives. However, we can follow precisely the route of chapter 6, and define the modified utilitarian problem with a constraint on the total amount of the resource to be consumed over time:

$$W(\beta) = \max \int_0^\infty u(c, s)e^{-\delta t} \, dt \quad \text{s.t.} \tag{9.9}$$

$$\dot{k} = F(k, \sigma) - c, \quad \int_0^\infty \sigma_t \, dt \leq \beta s_0, \quad 0 \leq \beta \leq 1$$

Here we are optimizing subject to the constraint that not more than a fraction β of the initial stock is consumed. Assume as before that the utility function is additively separable: $u(c, s) = u_1(c) + u_2(s)$. Let s^* be the stock level conserved forever in the discounted utilitarian solution, that is, a solution to (9.1). As in chapter 6, define β^* by

$$\beta^* s_0 = s_0 - s^*$$

so that it is the fraction of the stock consumed on a utilitarian optimal path with no constraint on consumption. Some basic properties of the valuation function $W(\beta s_0)$ are important and were established in chapter 6, and are restated here:

LEMMA 37 The valuation function $W(\beta s_0)$ is nondecreasing as a function of β for $\beta < \beta^*$, and is constant as a function of β for $\beta \geq \beta^*$.

β^* is the fraction of the initial stock consumed along a utilitarian optimum, and of course depends on the discount rate δ and the initial stock s_0. If we pick a consumption path to solve (9.9), then the total payoff from this path if we evaluate it according to the Chichilnisky criterion is given by the present value of utilities associated with this path, which is $\alpha W(\beta s_0)$, *plus* the limiting utility value, which is $(1 - \alpha) \lim_{t \to \infty} u(c_t, s_t)$. The expression for this depends on whether the amount of the initial stock made available for consumption, βs_0, is greater than or less than the amount consumed on a utilitarian optimal path, $\beta^* s_0$, that is, whether $\beta \geq \beta^*$. If $\beta < \beta^*$, the total payoff with the Chichilnisky criterion is

$$\alpha W(\beta s_0) + (1 - \alpha)u_2[(1 - \beta)s_0]$$

because in this case the constraint on cumulative consumption binds and the stock $(1 - \beta)s_0$ is maintained forever. If $\beta \geq \beta^*$, the cumulative consumption constraint is not binding, and the limiting stock is that associated with the utilitarian optimum. The total payoff is now

$$\alpha W(\beta) + (1 - \alpha)u_2(s^*)$$

The overall problem (9.8) can now be solved by picking β to maximize the sum of $\alpha W(\beta s_0)$ and either $(1 - \alpha)u_2[(1 - \beta)s_0]$ or $(1 - \alpha)u_2(s^*)$ as appropriate. The approach to solving this problem is illustrated in figure 9.2, which reproduces figure 6.2.

The interesting case is that in which the resource is necessary to production, so that $c^* = 0$; in this case the analysis follows very closely that of chapter 6 and in particular figure 6.2. Setting $\beta = 0$ corresponds to the green golden rule, because it selects the maximum value of the limiting term; setting $\beta = \beta^*$ corresponds to the utilitarian solution. In general the Chichilnisky criterion requires a value of β strictly between these two values, and hence a larger permanent stock than is selected by utilitarianism but a smaller one than implied by the green golden rule. Figure 9.3 compares, for the case of an essential resource, the three solutions discussed, according to utilitarian, limiting utility and

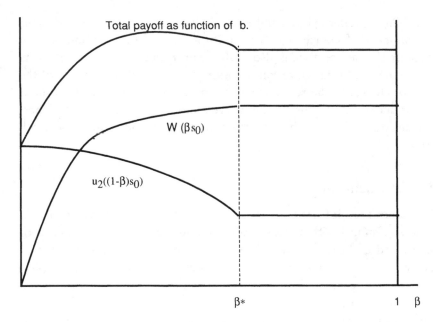

FIGURE 9.2   The optimal allocation of the initial stock between consumption and preservation.

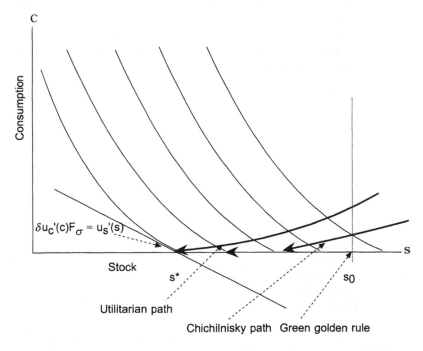

FIGURE 9.3   A comparison of alternative solution paths.

Chichilnisky criteria; it resembles closely the similar figure 6.3 of chapter 6, but now of course represents only a part of the dynamic system because the capital stock is also evolving over time and of course affects consumption. In this case, consumption is no longer the negative of the rate of change of the resource stock, and the resource flow is an input to production. Nevertheless, for a given initial capital stock, the higher-dimensional system can be projected into the c–s plane without too much loss.

9.4   Conclusion

The central result from this chapter is the following. We have previously studied a model with an exhaustible resource that can be consumed or conserved and provides utility both as a consumption good and as a stock of an environmental asset. For this model we obtained a complete and intuitive characterization of the paths that are optimal with respect to various criteria of optimality. Now we have extended the model substantially by bringing in production and capital accumulation; the resource is still valued as a stock, and has an alternative use as an input to production. This is the only use for the flow now; the flow cannot be consumed directly, but must be used as an input to production. The important point is that although this makes the technical details of the analysis more demanding, it does not in any qualitative way change the conclusions that emerge about the nature of optimal use patterns.

Chapter 10
Capital and Renewable Resources

Now we consider the most challenging of all cases, and perhaps most realistic and rewarding: an economy in which a resource that is renewable, and so has its own dynamics, can be used together with produced capital goods as an input to the production of an output. The output in turn can be reinvested in capital formation or consumed. The stock of the resource is also a source of utility to the population. So, using the notation of chapter 9, capital accumulation occurs according to

$$\dot{k} = F(k, \sigma) - c$$

and the resource stock evolves according to

$$\dot{s} = r(s) - \sigma$$

where k is the current capital stock, σ the rate of use of the resource in production, and $F(k, \sigma)$ the production function. As in chapter 4, $r(s)$ is a growth function for the renewable resource, indicating the rate of growth when the stock is s. We make the same assumptions about $r(s)$ as in chapter 4: $r(s)$ is taken to be strictly concave, twice differentiable, increasing up to a maximum, and then decreasing, and to satisfy $r(0) = 0$, and that there exists a positive $\bar{s}$ such that $r(\bar{s}) = 0$.

As before, we consider the optimum time path of resource use according to the utilitarian criterion, then characterize the green golden rule, the configuration of resources and capital that leads to the maximum sustainable utility level, and finally draw on the results of these two cases to characterize optimality according to Chichilnisky's criterion.

10.1 The Utilitarian Optimum

The utilitarian optimum in this framework is the solution to

$$\max \int_0^\infty u(c_t, s_t)e^{-\delta t}\, dt \quad \text{s.t.} \tag{10.1}$$

$$\dot{k} = F(k, \sigma) - c \quad \text{and} \quad \dot{s} = r(s) - \sigma$$

We proceed in the by-now standard manner, constructing the Hamiltonian

$$H = u(c, s)e^{-\delta t} + \lambda e^{-\delta t}[F(k, \sigma) - c] + \mu e^{-\delta t}[r(s) - \sigma]$$

and deriving the following conditions, which are necessary for a solution to (10.1):

$$u_c = \lambda \tag{10.2}$$

$$\lambda F_\sigma = \mu \tag{10.3}$$

$$\dot{\lambda} - \delta \lambda = -\lambda F_k \tag{10.4}$$

$$\dot{\mu} - \delta \mu = -u_s - \mu r_s \tag{10.5}$$

where r_s is the derivative of r with respect to the stock s.

10.1.1 Stationary Solutions — A little algebra shows that the system (10.2) to (10.5), together with the two underlying differential equations in (10.1), admits the following stationary solution:

$$\delta = F_k(k, \sigma) \tag{10.6}$$

$$\sigma = r(s) \tag{10.7}$$

$$c = F(k, \sigma) \tag{10.8}$$

$$\frac{u_s(c, s)}{u_c(c, s)} = F_\sigma(k, \sigma)(\delta - r_s) \tag{10.9}$$

This system of four equations suffices to determine the stationary values of the variables k, s, σ, and c. When r_s is zero and the resource is nonrenewable, this reduces to the equivalent system (9.6) and (9.7) of chapter 9. And the last equation here is an obvious generalization of the equivalent equation without production, which was $\delta - r_s = u_s(c, s)/u_c(c, s)$, noting that in that case $F_\sigma = 1$.

It is important to understand fully the structure of stationary states in this model, and in particular the trade-off between consumption c and the resource stock s across alternative stationary states. This is more complex than in any model we have considered so far.

First, consider this relationship across stationary states for a given value of the capital stock k: in a stationary state, $\sigma = r(s)$, so that in this case we can write

$$c = F[k, r(s)]$$

across stationary states. Hence, across stationary states the derivative of consumption c with respect to the stock of the environmental asset s for a fixed capital stock k is

$$\left.\frac{\partial c}{\partial s}\right|_{k \text{ fixed}} = F_\sigma r_s \qquad (10.10)$$

Because F_σ is always positive, this has the sign of r_s, which is initially positive and then switches to negative; hence, we have a single-peaked relationship between c and s for fixed k across stationary states. The c–s relationship across stationary states for a fixed value of k replicates the shape of the growth function $r(s)$ and so has a maximum for the same value of s.

We assume that the output level $F(k, \sigma)$ is an increasing function of the capital stock k for a given value of the environmental input σ. Hence, across stationary states characterized by different values of the environmental stock s, for each value of the stock s, output is increasing in the capital stock k. Hence, the c–s curves for fixed values of the capital stock k are ordered by the values of k, and those corresponding to higher k values dominate those corresponding to lower values of k, as shown in figure 10.1.

In general, however, k is not fixed across stationary states, but depends on σ via equation (10.6). Taking account of this dependence and treating (10.6) as an implicit function,

$$F_k(k, \sigma) - \delta = 0$$

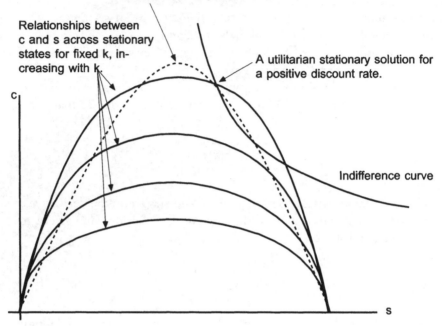

FIGURE 10.1   The relationship between consumption and the environmental stock across steady states.

we obtain the total derivative of c with respect to s across stationary states:

$$\left.\frac{dc}{ds}\right|_{k \text{ varies}} = F_k \frac{dk}{d\sigma}\frac{d\sigma}{ds} + F_\sigma \frac{d\sigma}{ds} = r_s\left(-\delta\frac{F_{k\sigma}}{F_{kk}} + F_\sigma\right) \quad (10.11)$$

which, maintaining the assumption that $F_{k\sigma} \geq 0$, also has the sign of r_s and again inherits the shape of $r(s)$. Note that for a given value of s

$$\left|\left.\frac{dc}{ds}\right|_{k \text{ varies}}\right| \geq \left|\left.\frac{\partial c}{\partial s}\right|_{k \text{ fixed}}\right|$$

so that the curve relating c to s across stationary states allowing k to adjust to bring the marginal product of capital into equality with the discount rate is steeper than that obtained when k remains fixed. The c–s curve on which k adjusts to bring F_k into equality with δ crosses the curve corresponding to a fixed value of k, say $\widehat{k}$, at the values of the environmental stock s, call them $\widehat{s}$, at which $F_k[\widehat{k}, r(\widehat{s})] = \delta$. At these stock values, the value of r is such that $F_k = \delta$ and the capital stock is fully adjusted to the discount rate. The two curves are equal only if the cross-derivative $F_{k\sigma}$ is zero.

The various curves relating c and s across stationary solutions are shown in figure 10.1; for $F_{k\sigma} > 0$ the curve corresponding to k fully adjusted to s rises and falls more sharply than the others, and crosses each of these twice, from below while increasing and from above while decreasing, as shown. A stationary solution with a capital stock $\widehat{k}$ must lie on the intersection of the c–s curve corresponding to a capital stock fixed at $\widehat{k}$ with the curve representing the fully adjusted relationship. At this point, c, s, and k are all fully adjusted to each other. (In the case of $F_{k\sigma} = 0$ the curves relating c and s for k fixed and fully adjusted are identical, so that in the case of a separable production function the dynamics are simpler, although qualitatively similar.)

The stationary first-order condition (10.9) relates most closely to the curve connecting c and s for a fixed value of k, and would indicate a tangency between this curve (whose slope is $F_\sigma r_s$) and an indifference curve if the discount rate δ were equal to zero. For positive δ, the case we are considering now, the stationary solution lies at the point where the c–s curve for the fixed value of k associated with the stationary solution crosses the c–s curve along which k varies with s. At this point, an indifference curve crosses the fixed-k c–s curve from above; this is shown in figure 10.1. Note that as we vary the discount rate δ, the capital stock associated with a stationary solution changes via the equation (10.6), so that in particular lowering the discount rate leads to a stationary solution on a fixed-k c–s curve corresponding to a larger value of k and therefore outside the curve corresponding to the initial lower discount rate.

10.1.2 Dynamics of the Utilitarian Solution — The four differential equations governing a utilitarian solution are

$$\dot{k} = F(k, \sigma) - c(s_t, \lambda_t)$$

$$\dot{s} = r(s) - \sigma(\mu_t, \lambda_t, k_t)$$

$$\dot{\lambda} - \delta\lambda = -\lambda F_k$$

$$\dot{\mu} - \delta\mu = -u_s - \mu r_s$$

The linearized system is

$$
\begin{bmatrix} \dot{k} \\ \dot{s} \\ \dot{\lambda} \\ \dot{\mu} \end{bmatrix} =
\begin{bmatrix}
\delta - F_\sigma \dfrac{F_{\sigma k}}{F_{\sigma\sigma}} & \dfrac{u_{cs}}{u_{cc}} & -\dfrac{F_\sigma F_\sigma}{\lambda F_{\sigma\sigma}} - \dfrac{1}{u_{cc}} & \dfrac{F_\sigma}{\lambda F_{\sigma\sigma}} \\[2ex]
\dfrac{F_{\sigma k}}{F_{\sigma\sigma}} & r_s & \dfrac{F_\sigma}{\lambda F_{\sigma\sigma}} & -\dfrac{1}{\lambda F_{\sigma\sigma}} \\[2ex]
-\lambda F_{kk} + \lambda F_{k\sigma}\dfrac{F_{\sigma k}}{F_{\sigma\sigma}} & 0 & \dfrac{F_{k\sigma}F_\sigma}{F_{\sigma\sigma}} & -\dfrac{F_{k\sigma}}{F_{\sigma\sigma}} \\[2ex]
0 & \dfrac{u_{sc}u_{cs}}{u_{cc}} - u_{ss} - \mu r_{ss} & -\dfrac{u_{sc}}{u_{cc}} & \delta - r_s
\end{bmatrix}
\begin{bmatrix} k \\ s \\ \lambda \\ \mu \end{bmatrix}
$$

To establish clear general results on the signs of the eigenvalues of the matrix of this system, as before we have to make simplifying assumptions. In fact, the assumptions used in the case of an exhaustible resource also suffice in this case. If $F_{\sigma\sigma}$ is large, so that the marginal productivity of the resource drops rapidly as more of it is used, then the eigenvalues of the above matrix are

$$ r_s, \qquad \delta - r_s, \qquad +\frac{\delta}{2} \pm \frac{1}{2u_{cc}}\sqrt{u_{cc}^2\delta^2 + 4u_{cc}\lambda F_{kk}} \qquad (10.12) $$

There are two negative roots in this case, as $r_s < 0$ at a stationary solution. In this case the utilitarian stationary solution is locally a saddle point.

PROPOSITION 38 A sufficient condition for the utilitarian stationary solution to be locally a saddlepoint is that $F_{\sigma\sigma}$ is large, so that the marginal productivity of the resource diminishes rapidly in production.

There are other cases in which the stationary solution is locally a saddlepoint. We leave those as an exercise to the reader.

10.2 The Green Golden Rule

Across stationary states, the relationship between consumption and the resource stock satisfies the equation

$$c = F[k, r(s)]$$

so that at the green golden rule we seek to maximize the sustainable utility level with respect to the inputs of capital k and the resource stock s:

$$\max_{s,k} u\{F[k, r(s)], s\}$$

Maximization with respect to the resource stock gives

$$\frac{u_s}{u_c} = -F_\sigma r_s \tag{10.13}$$

which is precisely the condition (10.9) characterizing the stationary solution to the utilitarian conditions for the case in which the discount rate δ is equal to zero. So, as before, the utilitarian solution with a zero discount rate meets the first-order conditions for maximization of sustainable utility with respect to the resource stock. Of course, in general the utilitarian problem may have no solution when the discount rate is zero. Note that condition (10.13) is quite intuitive and in keeping with earlier results. It requires that an indifference curve be tangent to the curve relating c to s across stationary states for k fixed at the level $\bar{k}$ defined below; in other words, it again requires equality of marginal rates of transformation and substitution between stocks and flows.

The capital stock k in the maximand here is independent of s. How is the capital stock chosen? In a utilitarian solution the discount rate plays a role in this through the equality of the marginal product of capital with the discount rate (10.6); this gives us the curve relating c to s for k fully adjusted in figure 10.1. Unfortunately, at the green golden rule there is no relationship equivalent to $F_k = \delta$ that can fix the capital stock.

I close the system in the present case by supposing that the production technology ultimately displays satiation with respect to the capital input alone; for each level of the resource input σ there is a level of capital stock at which the marginal product of capital is zero. Precisely,

$$\bar{k}(\sigma) = \min k: \frac{\partial F(k, \sigma)}{\partial k} = 0 \tag{10.14}$$

We call $\bar{k}(\sigma)$ the satiation capital stock and assume that $\bar{k}(\sigma)$ exists for all $\sigma \geq 0$ and is finite, continuous, and nondecreasing in σ. Essentially assumption (10.14) says that there is a limit to the extent to which capital can be substituted for resources: as we apply more and more capital to a fixed input of resources, output reaches a maximum above which it cannot be increased for that level of resource input. In the case in which the resource is an energy source, this assumption was shown by Berry, Heal, and Salomon [16] to be implied by the second law of thermodynamics; this issue is also discussed by Dasgupta and Heal [41]. In general, this seems to be a very mild and reasonable assumption.

Given assumption (10.14), the maximization of stationary utility with respect to the capital stock at a given resource input,

$$\max_k u\{F[k, r(s)], s\}$$

requires that we pick the capital stock at which satiation occurs at this resource input, i.e., $k = \bar{k}[r(s)]$.

Note that

$$\frac{\partial \bar{k}[r(s)]}{\partial s} = \frac{\partial \bar{k}(\sigma)}{\partial \sigma} r_s$$

so that the satiation capital stock $\bar{k}$ is increasing and then decreasing in the resource stock s across stationary states; the derivative has the sign of r_s.

We can now see that the green golden rule is the solution to the following problem:

$$\max_s u(F\{\bar{k}[r(s)], r(s)\}, s)$$

where at each value of the resource stock s the input of the resource and the capital stock are adjusted so that the resource stock is stationary and the capital stock maximizes output for that stationary resource input. The relationship between consumption and the resource stock across stationary states when at each resource stock the capital stock is adjusted to the level $\bar{k}[r(s)]$ satisfies the following equation:

$$c(s) = F\{\bar{k}[r(s)], r(s)\}$$

whose slope is

$$\frac{dc}{ds} = F_k\{\bar{k}[r(s)], r(s)\} \frac{\partial \bar{k}}{\partial \sigma} r_s(s) + F_\sigma\{\bar{k}[r(s)], r(s)\} r_s(s)$$

and by the definition of $\bar{k}$ the first term on the right is zero, so that the slope of the curve relating c and s when the capital stock is given by $\bar{k}$ is the same as the slope when the capital stock is fixed, namely,

$$\frac{dc}{ds} = F_\sigma\{\bar{k}[r(s)], r(s)\}r_s(s)$$

This curve shows the relationship between c and s when k adjusts fully, but the adjustment at issue here is not that of the previous section based on the equality of the marginal product of capital and the discount rate. Recall that by construction

$$F\{\bar{k}[r(s)], r(s)\} \geq F[k, r(s)]$$

for any value of the capital stock k. Hence this curve dominates all other c–s curves considered. The c–s curve corresponding to a particular k value, say $\hat{k}$, touches it at the s values, if any, for which $\bar{k}[r(s)] = \hat{k}$. In words, the c–s curve corresponding to a fixed value of k, $\hat{k}$, touches the c–s curve for k fully adjusted at environmental stocks s for which $\hat{k}$ is the satiation capital stock. The curve corresponding to full adjustment of k is thus the outer envelope of the curves for fixed values of the capital stock.

The total derivative of the utility level with respect to the stock of the resource is now

$$\frac{du}{ds} = u_c F_k \frac{\partial \bar{k}}{\partial \sigma} r_s + u_c F_\sigma r_s + u_s$$

By assumption (10.14) and the definition of $\bar{k}$, $F_k = 0$ here; hence, equating this to zero for a maximum sustainable utility level gives the earlier expression (10.13). The green golden rule is characterized by a tangency between an indifference curve and the outer envelope of all curves relating c to s across stationary states for fixed capital stocks.

PROPOSITION 39 In an economy with capital accumulation and renewable resources, under the assumption (10.14) of satiation of the production function with respect to capital, the green golden rule satisfies the first-order condition $u_s/u_c = -F_\sigma r_s$, which defines a tangency between an indifference curve and the outer envelope of c–s curves for fixed values of k. It has a capital stock of $\bar{k}[\sigma(s^*)]$, where s^* is the green golden rule value of the resource stock and $\bar{k}$ denotes the capital stock at which the marginal product of capital first becomes zero for a resource input of $\sigma(s^*)$.

What if production does not display satiation with respect to the capital stock? In this case there is no maximum to the output that can be obtained from a given resource flow and so from a given resource stock. Unless we assume satiation of preferences with respect to consumption, as in section 9.2, the green golden rule is not well defined. In that case we can generalize the analysis of section 9.2 to the present case and obtain similar results.

10.3   The Chichilnisky Criterion

Having studied the utilitarian and green golden rule approaches in the context of capital accumulation, production, and renewable resources, one task remains: the analysis of the implications of Chichilnisky's criterion in this context. This means solving the problem

$$\max \alpha \int_0^\infty u(c_t, s_t) e^{-\delta t} \, dt + (1 - \alpha) \lim_{t \to \infty} u(c_t, s_t) \tag{10.15}$$

s.t. $\dot{k} = F(k_t, \sigma_t) - c_t$ and $\dot{s}_t = r(s) - \sigma_t, \qquad s_t \geq 0 \, \forall t$

In the case of satiation of the production process with respect to capital, as captured by assumption (10.14), the situation resembles that with the Chichilnisky criterion with renewable resources in chapter 7: there is no solution unless the discount rate declines to zero. Formally,

PROPOSITION 40 Assume that condition (10.14) is satisfied, so that for any given input of the environmental resource the marginal productivity of capital falls to zero. Then problem (10.15) has no solution, that is, there is no optimum for the mixed optimality criterion.

PROOF The structure of the proof is the same as that used in chapter 7. The integral term is maximized by the utilitarian solution, which requires an asymptotic approach to the utilitarian stationary state. The limit term is maximized on any path that asymptotes to the green golden rule. Given any fraction $\beta \in (0, 1)$ we can find a path that attains the fraction β of the payoff to the utilitarian optimum and then approaches the green golden rule. This is true for any value of $\beta < 1$, but not true for

$\beta = 1$. Hence, any path can be dominated by another corresponding to a higher value of β.

We now consider instead optimization with respect to Chichilnisky's criterion with a discount rate that declines to zero over time.

PROPOSITION 41 Assume that condition (10.14) is satisfied, so that for any given input of the environmental resource the marginal productivity of capital falls to zero. Consider the problem

$$\max \alpha \int_0^\infty u(c, s)\, \Delta(t)\, dt + (1 - \alpha) \lim_{t \to \infty} u(c, s), \qquad 0 < \alpha < 1$$

$$\text{s.t.} \quad \dot{k} = F(k_t, \sigma_t) - c_t \quad \text{and} \quad \dot{s}_t = r(s_t) - c_t, \qquad s_0 \text{ given}$$

where $\int_0^\infty \Delta(t)\, dt < \infty$, $q(t) = -\dot{\Delta}(t)/\Delta(t)$ and $\lim_{t \to \infty} q(t) = 0$. This problem has a solution that is identical to that which maximizes $\int_0^\infty u_1(c, s)\, \Delta(t)\, dt$ subject to the same constraint. In words, solving the utilitarian problem with the variable discount rate that goes to zero solves the overall problem.

PROOF The proof is a straightforward adaptation of the proof of proposition 21 of chapter 7, and is omitted. The only point at which the proof differs is that one has now to show that the utilitarian solution approaches the green golden rule capital and resource stocks as the discount rate approaches zero. We have seen from section 10.2 that the utilitarian stationary solution for a zero discount rate satisfies the same first-order condition with respect to the resource stock as the green golden rule: compare (10.9) $u_s(c, s)/u_c(c, s) = F_\sigma(k, \sigma)(\delta - r_s)$ with $u_s/u_c = -F_\sigma r_s$ from proposition 39. At the green golden rule the marginal product of capital is zero as the capital stock is given by $\overline{k}[\sigma(s^*)]$; in the utilitarian stationary solution the marginal product equals the discount rate, which in the present case is zero. This completes the additional argument required.

What does the Chichilnisky-optimal path look like in this case? As figure 10.2 illustrates, the optimal path moves toward the green golden rule, which is a point of tangency between an indifference curve and the outer envelope of the curves relating c and s for fixed values of k. This point is the limit of utilitarian stationary solutions as the associated discount rate goes to zero. Note that from the expression (10.12), the green golden rule is a saddlepoint of the autonomous differential equation system describing the optimal path as the discount rate falls to zero.

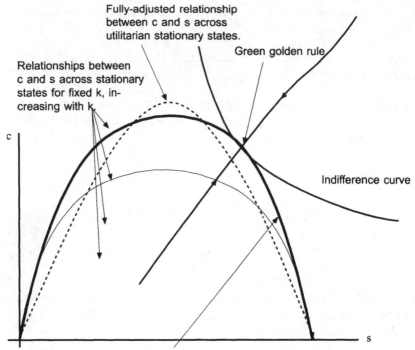

Fully-adjusted relationship
between c and s across
utilitarian stationary states.

Green golden rule

Relationships between
c and s across stationary
states for fixed k, in-
creasing with k

c

Indifference curve

Curve relating c to s when the capital stock is at its satiation
level: this is the upper boundary of the c-s curves for fixed
values of the capital stock.

s

FIGURE 10.2: The optimal path according to the Chichilnisky criterion and
the utilitarian criterion with the discount rate going to zero.

10.4 Conclusions

We have gradually worked our way from simple problems to problems
that by any standards are demandingly complex, from the Hotelling
problem, a nice application of isoperimetric techniques from the classic
calculus of variations, to the problems solved in this chapter involving
asymptotically autonomous systems and the optimization problems of the
previous chapters, which take us outside the range of control theory in its
conventional formats. Complexity has increased along two dimensions;
these dimensions are the nature of the technological constraints facing
the economy, and the type of objective considered in defining optimality.

In the dimension of constraints on the economy, we have progressed
from a pure integral constraint, $\int_0^\infty c_t \, dt \leq s_0$, via a framework for
modeling the allocation of investment to a backstop technology, to a
model where the resource has a life of its own via a renewal function and

production and capital accumulation are both possible: $\dot{k} = F(k_t, \sigma_t) - c_t$ and $\dot{s}_t = r(s_t) - c_t$. In the arena of objectives we have considered discounted utility integrals, limiting utility values, and mixtures of the two à la Chichilnisky. We have also remarked on the implications of the Rawlsian and overtaking definitions of optimality.

Despite the range of cases considered, two basic solution paradigms stand out. These are those presented in chapters 6 and 7, in which we revisited the pure depletion model and the renewable resource model using the Chichilnisky definition of optimality.

These two paradigms of chapters 6 and 7 are quite distinct: when the resource is exhaustible, the discounted utilitarian solution is at one end of a spectrum of possible solutions ranked by the amount of the initial stock of the resource that is conserved forever. The green golden rule and overtaking solutions are at the other end. The Chichilnisky case is intermediate.

With a renewable resource, matters are different: the discounted utilitarian and green golden rule solutions are well defined, but the Chichilnisky solution is well defined if and only if the discount rate in the integral term $\alpha \int_0^\infty u_1(c, s) \, \Delta(t) \, dt + (1 - \alpha) \lim_{t \to \infty} u_1(c, s)$ goes asymptotically to zero. In this case it bridges the two solutions; in fact, they merge. This raises a set of interesting and far-reaching questions about the way in which we approach intertemporal welfare economics and the evaluation of long-run benefits. The results of this chapter, and of chapter 9, fit into one of these paradigms: they confirm that the basic solution paradigms are robust to such changes in specification as the inclusion or exclusion of production and capital accumulation.

Part IV
Policy Issues

Chapter 11
Measuring National Income

11.1 Introduction

In this chapter and the next two, we reach the most important part of this study: we set out the implications of the analysis that has gone before for two important and related policy issues, which are both topical and enduring. These are the measurement of national income and the evaluation of environmental investment and conservation projects. This chapter and the next are devoted to presenting and studying two different approaches to measuring national income in the context of dynamic, resource-based economies.

There is widespread recognition that our current practices in the areas of national income accounting and project evaluation are inadequate for reflecting environmental concerns. The current conventions regarding national income accounting were developed in the 1940s and 1950s to provide a statistical framework for the implementation of Keynesian macroeconomics and demand management. They were developed to provide the data needed for these tasks. They have grown more sophisticated over the years, but are still dominated by the legacy of a social and intellectual environment in which the natural environment was not an agenda item.

The agenda underlying the dominant conventions of cost–benefit analysis was also set several decades ago, in the 1960s and 1970s, when the central issues facing developing countries were balance of payments constraints, the use of domestic versus world prices, unemployment, but never the preservation of environmental assets, the evaluation of their contribution to society, or the evaluation of very long-term projects. It is not that any of the prior analyses of any of these topics were wrong; it is just that they were conducted in a framework that did not address core environmental concerns. They therefore must be supplemented.

11.2   Two Concepts of National Income

There are two distinct, although related, ways in which the term *national income* is used in the recent literature on national income and the environment. Underlying one usage is the concept of income widely attributed to Hicks, but also (as noted in chapter 1) found in the work of Fisher and Lindahl.[1] Hicks [63] stated that "Income No. 3 must be defined as the maximum amount of money which an individual can spend this week, and still expect to be able to spend the same amount in real terms in each ensuing week" (p. 174). This is often paraphrased as the maximum consumption that maintains capital intact. Hicks went on to introduce the concept of a "standard stream (of income) which is constant in real terms" and said, "We ask . . . how much he would be receiving if he were getting a standard stream of the same present value as his actual expected receipts. This amount is his income" (p. 184). So income is defined here as the expenditure that if kept constant would yield the same present value as a person's actual future receipts. We commented before that there is an implicit concept of sustainability in this definition. We refer to this concept from now on as the *Hicksian*[2] definition of income. Hicks proposed this as a definition of an individual's income, but the concept has quite naturally been applied to the income of a society, and has become an interpretation of national income. This definition has the advantage of being explicitly dynamic and addressing the passage of time. The first formalization of this interpretation of national income is due to Weitzman [109]; subsequently the concept has been developed further by Asheim [5], [8], Hartwick [55], and others, including Dasgupta [39] and Dasgupta, Kriström, and Mäler [44].

The alternative usage of *national income* is rooted in traditional welfare economics, and is an attempt to construct an index number with welfare significance. It is best explained geometrically.

Figure 11.1 shows an economy with two goods, labor and consumption. Labor can be converted into consumption according to the

---

[1] I am very much indebted to Geir Asheim for a most instructive discussion of Hicks's concept of income: his two papers [5] and [8] develop this concept in several interesting ways.

[2] In using this adjective I am conforming to widespread usage, nothwithstanding the fact that Fisher and Lindahl have claims to priority on some part of this concept.

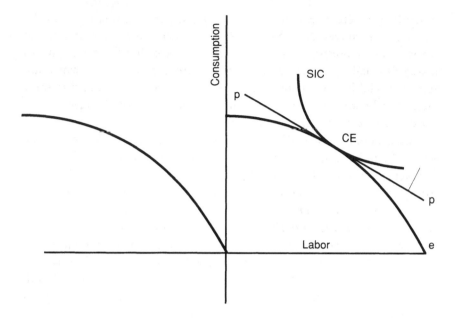

FIGURE 11.1   The standard welfare economics concept of national income.

production function shown to the left of the vertical axis (inputs are negative, outputs positive), the economy's endowment is exclusively of labor, at point *e*, and the transformation frontier showing alternative combinations of labor and consumption available to the economy is shown on the right of the vertical axis, as the production frontier transformed by the endowment vector. Point CE is a competitive equilibrium, with SIC being the social indifference curve corresponding to the individual indifference curves attained at the equilibrium.[3] Line *pp* is a hyperplane separating the set bounded by the social indifference curve from the feasible set. The normal to this line represents the equilibrium price vector: national income is now defined as the value of equilibrium consumption at these prices. It can be measured as the intercept of the separating hyperplane on either of the axes, or as the inner product. Note that this clearly has local welfare significance: any small move from CE that has a positive value at the prices defined by *pp* moves the economy above the social indifference curve through CE and so is

---

[3]This social indifference curve is the lower boundary of the set-theoretic sum of the individual preferred-or-indifferent sets corresponding to equilibrium consumption levels. For more discussion and a review of the literature on this topic, see Chichilnisky and Heal [26].

potentially Pareto improving. As the diagram shows, this need not be true for large moves. No significance is attached to the value of national income in this interpretation; the key thing here is that the prices give us a way of deciding which changes are welfare improving and which are not. National income is interesting because a necessary condition for a change to be improving (and a sufficient one for small changes) is that the change increase national income.

The generalization of figure 11.1 is national income defined as the value of equilibrium consumption at equilibrium prices, with the test of the welfare impact of a small change in the economy being whether it increases this measure. This is the concept of national income underlying cost–benefit analysis. This interpretation in national income we call the *national welfare interpretation,* or just national welfare.

In principle, economists have long been aware of the limitations of national welfare as an indicator of a nation's well-being. With respect to environmental matters, the shortcomings were clearly set out by Nordhaus and Tobin [88] many years back. In fact, even without the complexities introduced by environmental considerations, national income is at best a local measure of national well-being, and then only under rather special circumstances, as any introductory text on welfare economics will show (e.g., Graaff [50]). By local I mean that small changes that are increases in national income can be taken to be improvements in welfare, in the Paretian sense that if gainers compensate losers they could make someone better off and no one worse off. The same is not necessarily true of large changes: national income is only a first-order approximation to the "true" measure of national income.

In this chapter and the next I investigate what is the appropriate first-order measure of economic welfare in a dynamic economy with a resource base that is depleted for consumption, possibly renewed, and also enjoyed as a source of utility, as in the models of the previous chapters. I study the two approaches to national income, Hicksian income and national welfare. There is a body of literature that relates the Hamiltonian of a dynamic optimization problem to the Hicksian interpretation of national income. In particular, a linear approximation to the Hamiltonian of a dynamic economy can be related to the Hicksian definition of income for a society.

11.3   A General Model

To begin our analysis of national income concepts, we consider first a very general formulation of the optimal dynamics use of resources, of which all the problems in the previous chapters are special cases. Let the vector $c_t \in \mathcal{R}^m$ be a vector of flows of goods consumed and giving utility at time t, and $s_t \in \mathcal{R}^n$ be a vector of stocks at time t, also possibly, but not necessarily, sources of utility. Each stock $s_{i,t}$, $i = 1, \ldots, n$, changes over time in a way that depends on the values of all stocks and all flows:

$$\dot{s}_{i,t} = d_i(c_t, s_t), i = 1, \ldots, n \tag{11.1}$$

The economy's objective is to maximize the discounted integral of utilities:

$$\max \int_0^\infty u(c_t, s_t) e^{-\delta t} \, dt \tag{11.2}$$

subject to the rate-of-change equations (11.1) for the stocks. The utility function u is assumed to be strictly concave and the reproduction functions $d_i(c_t, s_t)$ are assumed to be concave.

To solve this problem we construct a Hamiltonian that takes the form

$$H_t = u(c_t, s_t) e^{-\delta t} + \sum_{i=1}^n \lambda_{i,t} e^{-\delta t} d_i(c_t, s_t) \tag{11.3}$$

where the $\lambda_{i,t}$ are the shadow prices of the stocks. The first-order conditions for optimality can be summarized as

$$\frac{\partial u(c_t, s_t)}{\partial c_j} = - \sum_{i=1}^n \lambda_{i,t} \frac{\partial d_i(c_t, s_t)}{\partial c_j} \tag{11.4}$$

and

$$\dot{\lambda}_{i,t} - \delta \lambda_{i,t} = - \frac{\partial u(c_t, s_t)}{\partial s_i} - \sum_{k=1}^n \lambda_{k,t} \frac{\partial d_k(c_t, s_t)}{\partial s_i} \tag{11.5}$$

Within this framework, we now investigate the two alternative approaches to defining national income. We turn first to the Hicksian interpretation of national income.

11.4   Hicksian Income and the Hamiltonian

11.4.1 Constant Discount Rates — In this section we establish a rather surprising connection between the Hamiltonian in a dynamic optimization problem and the Hicksian concept of national income. To be precise, we show that the value of the Hamiltonian at date t (not discounted back to date zero) represents a utility level that, if maintained forever starting at date t, would give the same present value of utility as the present value of utility along an optimal path from date t on. To formalize this point, let (c_t^*, s_t^*) be the solution to the problem of maximizing (11.2) subject to (11.1), as set out above. Also let CH_t be the Hamiltonian corresponding to this problem, not discounted to time zero; CH_t is the current value Hamiltonian. Then the following is true:

PROPOSITION 42 For any date t,

$$\int_t^\infty CH_t(c_t^*, s_t^*) e^{-\delta(\tau - t)} \, d\tau = \int_t^\infty u(c_\tau^*, s_\tau^*) e^{-\delta(\tau - t)} \, d\tau$$

A utility stream from date t to infinity of a constant value equal to the Hamiltonian evaluated on the optimal path at date t has the same present value as the utility stream from date t to infinity associated with a solution to the problem of maximizing (11.2) subject to (11.1).

PROOF Introduce a variable W_t, which is the utility level that, if maintained from t to infinity, would have the same present value as $(c_t^*, s_t^*)_{t,\infty}$:

$$\int_t^\infty W_t e^{-\delta(\tau - t)} \, d\tau = \frac{W_t}{\delta} = \int_t^\infty u(c_\tau^*, s_\tau^*) e^{-\delta(\tau - t)} \, d\tau$$

Obviously we need to show that $W_t = H_t$; we do this by showing that both satisfy the same differential equation. Clearly

$$\frac{d W_t}{dt} = \delta[-u(c_t^*, s_t^*) + W_t] \tag{11.6}$$

so

$$W_t = u(c_t^*, s_t^*) + \frac{1}{\delta} \frac{d W_t}{dt}$$

Now turn to the Hamiltonian

$$\frac{d CH_t}{dt} = \sum_i \frac{\partial u}{\partial c_i} \frac{dc_i}{dt} + \sum_i \frac{\partial u}{\partial s_i} \frac{ds_i}{dt} + \sum_i \frac{d\lambda_{i,t}}{dt} \frac{ds_i}{dt} + \sum_i \lambda_{i,t} \frac{d^2 s_i}{dt^2}$$

and, by simplifying and using the first-order conditions (11.4) and (11.5), this reduces to

$$\frac{d\,\mathrm{CH}_t}{dt} = \delta \sum_i \frac{ds_i}{dt} \lambda_{i,t} \tag{11.7}$$

Recall that

$$\mathrm{CH}_t = u(c_t^*, s_t^*) + \sum_{i=1}^{n} \lambda_{i,t} \frac{ds_i}{dt}$$

so that CH_t satisfies the differential equation

$$\mathrm{CH}_t = u(c_t^*, s_t^*) + \frac{1}{\delta} \frac{d\,\mathrm{CH}_t}{dt}$$

from which the desired result follows.

The Hamiltonian is thus a measure of the weighted average utility level associated with an optimal path with the weights being the discount factors. It is a natural candidate for a measure of Hicksian national income. In fact it is more convenient to work with a first-order approximation to the Hamiltonian. The point is that the absolute value of the Hamiltonian has no significance; what matters operationally is to know whether a given policy increases or decreases sustainable utility or Hicksian income. A test of this for policies that involve small changes is to see whether the changes in the variables implied by the policy have a positive inner product with the derivatives of the Hamiltonian with respect to those variables, that is, whether the policy increases the value of a linear approximation to the Hamiltonian.

Equation (11.7), derived as a step in the proof, is actually of interest in its own right. It tells us that the rate of change of the current value Hamiltonian equals the rate of accumulation of stocks valued at their shadow prices. Recognizing that the Hamiltonian is a measure of average future utility, we see that the change in the value of average future utility equals the accumulation of stocks multiplied by their shadow prices.

REMARK 1 The rate of change of weighted average future utility in the economy of the previous section is given by the rate of accumulation of stocks valued at their shadow prices.

Expression (11.6) is also of interest: it states that the rate of change of the equivalent constant utility level is equal to the difference between this level and the actual utility level, multiplied by the discount rate. This is a rather intuitive relationship.[4]

The form of the Hamiltonian gives us information about the structure of Hicksian income. It tells us that this should contain a function of the flows that affect utility and a measure of changes in wealth: $CH_t = u(c_t^*, s_t^*) + \sum_{i=1}^{n} \lambda_{i,t}(ds_i/dt)$. Note for example that capital gains on the stocks due to price changes do not appear here. Later in this chapter we study the linear approximation to the Hamiltonian, and show that this can be simplified in some informative ways. But first we study the extent to which our present results can be extended to the more general case of time-varying discount rates.

11.4.2 Time-Varying Discount Rates

The results we have just established do not carry over to the case of nonconstant discount rates. To see this, consider a problem identical to that of the previous two sections, except that the maximand is now

$$\max \int_0^\infty u(c_t, s_t)\, \Delta(t)\, dt \qquad (11.8)$$

where as usual we require that $\Delta(t)$ satisfy the conditions $\int_0^\infty \Delta(t)\, dt < \infty$ and $q(t) \equiv -\dot\Delta(t)/\Delta(t) \to 0$ as $t \to \infty$. In parallel to the previous analysis, define

$$\int_t^\infty W_t \Delta(\tau - t)\, d\tau = W_t \Gamma(t) = \int_t^\infty u(c_\tau^*, s_\tau^*)\, \Delta(\tau - t)\, d\tau \qquad (11.9)$$

where $\Gamma(t) = \int_t^\infty \Delta(\tau - t)\, d\tau$. Differentiating (11.9) with respect to time gives

$$W_t = u(c_t^*, s_t^*) + \frac{dW}{dt}\frac{1}{q_\infty(t)} \qquad (11.10)$$

where $\Delta(0)/\Gamma(t) = q_\infty(t)$; this can be interpreted as a long-term discount rate,[5] $q(t) = \dot\Delta/\Delta$ being the instantaneous discount rate. Both are equal to the discount rate δ in the case of exponential discounting $e^{-\delta t}$.

---

[4]This interpretation is developed more generally by Asheim [8].
[5]See Asheim [8] for a more detailed discussion.

If we use the obvious formulation of the Hamiltonian, namely,

$$H_t = u(c_t, s_t)\,\Delta(t) + \sum_{i=1}^{n} \lambda_{i,t}\,\Delta(t)\,d_i(c_t, s_t)$$

then it does not satisfy an equation of the same form as (11.10); it instead satisfies

$$CH_t = u(c_t^*, s_t^*) + \frac{d\,CH_t}{dt}\,\frac{1}{q(t)}$$

where $CH_t = u(c_t, s_t) + \sum_{i=1}^{n} \lambda_{i,t}\, d_i(c_t, s_t)$. Because CH and W do not satisfy the same equation, they are not identical.[6] The neat interpretation of the Hamiltonian as a measure of average future utility is therefore restricted to the case of constant discount rates.

11.5   The Linearized Hamiltonian

We next study a linear approximation to the current value Hamiltonian, which we call LH; as we have seen, this can be interpreted as a Hicksian measure of national income (and not as the more conventional measure,

---

[6]If instead of the Hamiltonian we use the following expression for national income (where $\Gamma(t)$ is as defined in the text)

$$\hat{H}_t = u(c_t, s_t)\Gamma(t) + \sum_{i=1}^{n} \lambda_{i,t}\Gamma(t)\,d_i(c_t, s_t)$$

and in addition we use the first-order conditions corresponding to this, namely,

$$\dot{\lambda}_{i,t} + \frac{\dot{\Gamma}(t)}{\Gamma(t)}\lambda_{i,t} = -\frac{\partial u(c_t, s_t)}{\partial s_i} - \sum_{k=1}^{n} \lambda_{k,t}\frac{\partial d_k(c_t, s_t)}{\partial s_i}$$

then we can verify that this implies that $CH_t = u(c_t^*, s_t^*) + \sum_{i=1}^{n} \lambda_{i,t}(ds_i/dt)$ satisfies

$$CH_t = u(c_t^*, s_t^*) + \frac{d\,CH_t}{dt}\,\frac{1}{q_x(t)}$$

In this formulation, instead of the Hamiltonian we have used an expression for national income in which the discount factor $\Delta(t)$ has been replaced by $\Gamma(t) = \int_t^x \Delta(\tau - t)\,d\tau$. One can interpret this as an average discount factor over the balance of the horizon, a long-term discount factor. So we have an equivalence, but not a very satisfactory one: it involves using different discount rates in calculating first-order conditions from those in the social optimization problem.

which is an index based on the prices supporting a separating hyperplane).

$$\text{LH} = \sum_{j=1}^{m} \frac{\partial u}{\partial c_j} c_j + \sum_{i=1}^{n} \frac{\partial u}{\partial s_j} s_j + \sum_{i=1}^{n} \lambda_{i,t} \left(\sum_{j=1}^{m} \frac{\partial d_i}{\partial c_j} c_j + \sum_{k=1}^{n} \frac{\partial d_i}{\partial s_k} s_k \right)$$

$$= \sum_{j=1}^{m} c_j \left(\frac{\partial u}{\partial c_j} + \sum_{i=1}^{n} \lambda_{i,t} \frac{\partial d_i}{\partial c_j} \right) + \sum_{i=1}^{n} s_i \left(\frac{\partial u}{\partial s_i} + \sum_{k=1}^{n} \lambda_{k,t} \frac{\partial d_k}{\partial s_i} \right)$$

$$(11.11)$$

The expression (11.11) indicates why the linearized Hamiltonian has interesting welfare properties. It multiplies all quantities in the economy—flows $c_j, j = 1, \ldots, m$ and stocks $s_i, i = 1, \ldots, n$—by their social values, which are their marginal contributions to utility and their marginal contributions to the growth of the stocks, where these stocks are valued at the shadow prices $\lambda_{k,t}$. So the term $\sum_{j=1}^{m} c_j[(\partial u/\partial c_j) + \sum_{i=1}^{n} \lambda_{i,t}(\partial d_i/\partial c_j)]$ represents the sum of consumption flows multiplied by their marginal utilities $(\partial u/\partial c_j)$ and by the opportunity costs of consumption measured in terms of the impact of increased consumption on the rates of growth of the stocks $[\sum_{i=1}^{n} \lambda_{i,t}(\partial d_i/\partial c_j)$; some of these terms may be negative]. Likewise, the term $\sum_{i=1}^{n} s_i[(\partial u/\partial s_i) + \sum_{k=1}^{n} \lambda_{k,t}(\partial d_k/\partial s_i)]$ represents a similar treatment of stocks. These multiples are then summed to give the aggregate social value of all stocks and flows.

The shadow prices used to value stocks of course depend on the objective chosen, as we have seen in earlier chapters and will see in more detail later in this chapter; this implies that Hicksian national income depends on the specification of the economy's objective, and in particular the relative weights assigned to different generations.

We are now in a position to establish a simple but fundamental relationship between the linearized Hamiltonian, the Hicksian national income, and the values of the stocks in the economy and the returns available on them. This relationship shows that the linearized Hamiltonian is just the real return on the nation's stocks. We see that the general expression for the linearized Hamiltonian is the sum of the shadow values of stocks each multiplied by a return that equals the extent to which price movements depart from the Hotelling rule on that stock. (The Hotelling rule is that price changes at the discount rate; the departure from this is the difference between price changes and the discount rate.) As we see below, this difference is the real rate of return on the stock.

PROPOSITION 43 Consider an economy whose operation is described by the maximization of $\int_0^\infty u(c_t, s_t)e^{-\delta t}\, dt$ subject to the constraints $\dot{s}_{i,t} = d_i(c_t, s_t)$, $i = 1, \ldots, n$. Assume the derivative of the Hamiltonian with respect to each stock s_i is nonzero on an optimal path. Then the linearized current value Hamiltonian is a return on the economy's stocks; it is equal to the value of the stocks in the economy at time t, valued at the shadow prices at time t, multiplied by the discount rate minus the rate of appreciation of the shadow prices at time t. Formally,

$$\mathrm{NI}_t = \sum_{i=1}^{n} s_{i,t} \lambda_{i,t} \left(\delta - \frac{\dot{\lambda}_{i,t}}{\lambda_{i,t}} \right)$$

PROOF Consider the linearized Hamiltonian LH (11.11). Note that in the second expression for LH, the first term in parentheses is equal to zero by the first-order conditions for optimality (11.4).[7] Consider now the second term:

$$\sum_{i=1}^{n} s_i \left(\frac{\partial u}{\partial s_i} + \sum_{k=1}^{n} \lambda_{k,t} \frac{\partial d_k}{\partial s_i} \right)$$

Then by condition (11.5) determining the rate of change of the shadow prices, this can be rewritten as

$$\sum_{i=1}^{n} s_i \left(\delta \lambda_{i,t} - \dot{\lambda}_{i,t} \right) = \sum_{i=1}^{n} s_i \lambda_{i,t} \left(\delta - \frac{\dot{\lambda}_{i,t}}{\lambda_{i,t}} \right)$$

which proves the proposition.

This simple result develops further the connection between Hicksian income and the Hamiltonian. It shows that the linearized Hamiltonian is equivalent to the Hicksian concept of income as a return on stocks, with a generalization that deals with non–steady-state behavior. Consumption flows are not featured in this expression. This result is very general, as the formulation of maximizing (11.2) subject to (11.1) is very general indeed, and encompasses all the utilitarian formulations considered so far, plus the conventional neoclassical formulations of optimal growth without

---

[7]Note that this is true even if we impose nonnegativity constraints on the flows c_t and consider the corresponding complementary slackness conditions.

natural resources. It also applies to the optimal resource depletion models considered by Hotelling and subsequently by Dasgupta and Heal and others. Note that in the case of a Hotelling model, because the rate of growth of the shadow price is equal to the discount rate, this measure of national income is always zero along an optimal path, as noted by Dasgupta and Heal [41]; more on this below.

The computation of the rate of return to be applied to the value of a stock is a key feature of this result; this is given by the discount rate minus the rate of change of the shadow prices of these stocks. This of course equals the discount rate in a stationary state, when by definition all shadow prices are constant. Generally, however, outside a stationary state the rate of return on the *i*th stock is the discount rate *minus* the rate of capital gain on this stock. This is not the rate of return to holding the stock, which would be the rental rate *plus* the capital gain, nor is it the total return with capital gains subtracted out. One can interpret the adjusted return used in the computation of Hicksian national income in two ways.

One interpretation is as the rate of discount expressed in terms of the ability to buy the *j*th stock, that is, deflated by the change in price of that stock. This is sometimes called the "own rate of discount" on the stock. The other interpretation is as the real rate of return on the stock. Consider the expression for the change in the shadow price, which from equation (11.5) is given by

$$\frac{\dot{\lambda}_{i,t}}{\lambda_{i,t}} + \frac{1}{\lambda_{i,t}}\left[\frac{\partial u(c_t, s_t)}{\partial s_i} + \sum_{k=1}^{n}\lambda_{k,t}\frac{\partial d_k(c_t, s_t)}{\partial s_i}\right] = \delta \qquad (11.12)$$

The left-hand side here is the total return on a unit of the *i*th stock; the first term is the capital gain and the second represents the contribution made by an extra unit of the stock to utility and to the growth of all stocks, multiplied by their shadow prices. This second term is the real return; these terms represent the return to an increment of the stock used in the economy. In the long run, with full adjustment of stocks to their appropriate levels, one would expect the real return to equal the discount rate. In a stationary state they are equal. The return applied to the value of stocks at each point in time, $\delta - \dot{\lambda}_{i,t}/\lambda_{i,t}$, equals the real return; at a stationary state this equals the discount rate. The key point is that if the discount rate exceeds the real return, that is $\dot{\lambda}_{i,t}/\lambda_{i,t}$ is positive, then $\sum_{i=1}^{n} s_{i,t}\lambda_{i,t}\delta$ exceeds the real return available on the economy's stocks and this has to be adjusted downward by an amount that depends on the difference between the two returns. The result in proposition 43 is therefore in the spirit of, but a generalization of, the conventional definition of national income descending from Fisher, Lindahl, and Hicks and discussed in chapter 1; recall that this definition of income states it to be the maximum amount that can be consumed without reducing capital, that is, the return on capital.

11.5.1 Consumption and Hicksian National Income — It is important to understand the role of consumption flows in Hicksian national income. Although they appear in the Hamiltonian, they are netted out in the linearization and simplification carried out above. This is not because they are not important—quite the contrary. It is because the levels of these flows are chosen to maximize the Hamiltonian insofar as it depends on them, in which case the derivative of the Hamiltonian with respect to the flows is zero because it is evaluated at a turning point with respect to these variables. Expression (11.4) indicates that the benefits of an increase in a consumption flow are just offset on the margin by the costs in terms of reduced accumulation of stocks; this is exactly how matters should be. Consumption has a benefit, the marginal utility of consumption, and a cost, in terms of the depletion of a resource if the resource is exhaustible, and otherwise in terms of reduced accumulation of the resource, with both leading to lower consumption levels in the future. Of course, this does not imply that more consumption would not be valuable; it just implies that within the constraints of the system a change in consumption cannot improve welfare. In particular, it implies that for exhaustible resources, the consumption of the resource must be associated with a term in the expression for Hicksian national income that reflects the associated depletion of the asset that is the resource stock.

If a unit of consumption were to be donated to the economy from outside, at no cost in terms of accumulation of stocks, it would of course increase national income as it would be multiplied by the term $\partial u(c_t, s_t)/\partial c_j$, which is nonzero. In summary, along an optimal path a variable that is chosen so as to maximize the Hamiltonian will not appear in a linear approximation to the Hamiltonian, more or less by definition, but a gift of a unit of that variable from outside the economy will typically still be valued. This explains why in the Hotelling model the measure of Hicksian national income along an optimal path is always zero; this does not imply that there is no welfare associated with such a path. The alternative welfare and national income measure to be considered below does assign a positive national income level to an optimal path in the Hotelling model. As remarked earlier, the level of the national income measure is arbitrary. Significance attaches only to the direction in which it changes.

In the alternative interpretation of national income, as national welfare or a generalization of the separating hyperplane approach used in general equilibrium theory, we shall see that in contrast to the results above, national income does depend on consumption levels, and there is no term to reflect depreciation of the economy's resource base.

11.5.2 Hicksian Income and the Discount Rate — A noteworthy aspect of the result in proposition 43 is the way in which the measure of Hicksian national income depends on the discount rate. For a fixed set of stocks and price paths, Hicksian national income appears to decline as the discount rate is lowered. In fact, all these variables are related; in a stationary state of most of the models considered, the stocks depend on the discount rate and increase with a decrease in the discount rate. So across stationary states, it is possible that a small decrease in the discount rate could raise national income. For any given path, of course, there is only one stationary state, so that a comparison across stationary states is of limited relevance.

In general, there is difficulty in interpreting this concept of national income when the discount rate is zero. If the discount rate is always zero, the integrals used in defining Hicksian income, $\int_t^\infty H_t(c_t^*, s_t^*) e^{-\delta(\tau - t)} \, d\tau$ and $\int_t^\infty u(c_t^*, s_t^*) e^{-\delta(\tau - t)} \, d\tau$, are not well defined, so that the equivalence of the Hamiltonian and Hicksian income cannot be established.

11.6 Hicksian Income and Resources

In this section we study the application of the results of the previous section on Hicksian national income to resource-based economies, first to economies with exhaustible resources, then to those with renewable resources, and finally to those with capital accumulation and renewable resources. We see the exact implications of this approach to defining national income in these contexts.

11.6.1 Exhaustible Resources — We now set out in detail the implications of proposition 43 to the models considered in earlier chapters. For the model of depletion of an exhaustible resource, the Hamiltonian for the utilitarian framework is

$$H_t = u_1(c_t) + u_2(s_t) - \lambda_t c_t$$

and using the first-order condition for maximization of the Hamiltonian with respect to the level of consumption, a linear approximation is[8]

$$\text{HNI}_t = c_t u_1'(c_t) + s_t u_2'(s_t) - c_t u_1'(c_t)$$

$$\text{i.e.,} \quad \text{HNI}_t = s_t u_2'(s_t)$$

where HNI stands for Hicksian national income. So consumption flows net out and the value of HNI at t is just the present value of the flow of services from the resource stock, valued at the marginal utility of the stock. We know that at a stationary solution s^* the marginal utility satisfies

$$u_2' = \delta u_1'$$

With this relationship we can rewrite NI at a stationary state s^* as

$$\text{HNI}_t = \delta s^* \lambda_t \tag{11.13}$$

which is just the shadow value of the stock multiplied by the discount rate, as indicated by proposition 43. This is, of course, the traditional definition of income: the flow of services from a capital stock. Away from a stationary state, the corresponding equation is

$$\text{HNI}_t = \left(\delta - \frac{\dot{\lambda}}{\lambda}\right)\lambda_t s_t \tag{11.14}$$

As noted earlier, the flow of utility from depleting the resource makes no contribution to HNI; the value of the flow $c_t u_1'(c_t)$ is exactly offset by a term $\lambda_t c_t$ accounting for the depletion of the stock. Only the stock counts: if the stock were not valued in our economy, then HNI would be zero. Any change in the stock, through discoveries or through sales, must be recorded in HNI and valued (in a stationary state) at the shadow price of the flow times the discount rate. The discount rate affects the measure

---

[8]I remove the discount factors $\exp(-\delta t)$ to put the Hamiltonian and national income in current value terms at time t, not in present value terms.

of HNI directly, as already noted several times. As already remarked, the absence of consumption in the Hicksian national income expression does not imply that a gift of consumption from outside the economy is valueless; such a gift would be valued at the rate $u_1'c_t = \lambda_t > 0$. Also, it does not imply that a loss of stock has no cost: this would have a cost of the same amount.

If we use the Chichilnisky criterion rather than the utilitarian as a maximand, then exactly the same principles apply, except that the shadow price of consumption and the stock in the stationary state are higher. In this case, the economy does not reach the stationary state at which $u_2' = \delta u_1'$. The stationary state now satisfies (6.4), which is repeated here:

$$\frac{u_2'(\hat{s})}{u_1'(0)} = \frac{\alpha\delta}{(1-\alpha)\delta + \alpha}$$

so that in a stationary state

$$\text{HNI}_t = \hat{s}u_1'(0)\frac{\alpha\delta}{(1-\alpha)\delta + \alpha}$$

which reduces to the previous expression for the utilitarian case when $\alpha = 1$. Note that $\hat{s}$ is a function of α and increases as α decreases. The expression $\alpha\delta/[(1-\alpha)\delta + \alpha]$ decreases as α decreases, so that it is not generally possible to say how the stationary state HNI changes with an increase in the weight put on the long-run utility level. The expression for HNI outside the stationary state, (11.14), still applies in the case of Chichilnisky's criterion, although of course the shadow prices are different.

Recall that at each point in time, the shadow price of the resource is equal to the marginal utility of consumption, from equation (4.3) of chapter 4. In this case, we have from equation (6.3) of chapter 6 that

$$-\alpha u_1'(0) + \frac{\alpha u_2'(\hat{s})}{\delta} + (1-\alpha)u_2'(\hat{s}) = 0$$

or

$$\hat{\lambda} = \frac{u_2'(\hat{s})}{\delta} + \frac{u_2'(\hat{s})(1-\alpha)}{\alpha}$$

where $\hat{\lambda}$ is the shadow price of the resource when the stationary state is reached. Here the second term in the expression for the shadow price represents the contribution of an increase in the stock to the limiting utility level; the first term of course represents its contribution to the discounted sum of utilities. The second term, although it represents contributions in the far distant future, is not discounted. The move from a utilitarian to a Chichilnisky objective function has altered the stationary shadow price, and has introduced into this a term representing the undiscounted value of the impact of the stock on long-run utility levels.

11.6.2 Renewable Resources — Similar results hold in the case of renewable resources. The Hamiltonian is now

$$H_t = [u_1(c_t) + u_2(s_t)] + \lambda_t[r(s_t) - c_t]$$

so that

$$\text{HNI}_t = [c_t u_1'(c_t) + s_t u_2'(s_t)] + u_1'[s_t r'(s_t) - c_t]$$

As before, the terms in the flow c_t net out, so that

$$\text{HNI}_t = s_t[u_2'(s_t) + u_1' r'(s_t)] = s_t \lambda_t \left(\delta - \frac{\dot{\lambda}_t}{\lambda_t} \right)$$

and

$$\text{HNI}_t = \delta \lambda_t s_t$$

in a stationary solution. Once again, only terms relating to the stock appear in the expression for HNI. Recall that r' may be positive or negative; near the green golden rule, it is negative.

11.6.3 National Income and Capital Accumulation — Finally, we investigate whether the introduction of production and capital accumulation alters matters at all. Not surprisingly, the answer is that nothing essential is changed: all the principles established so far continue to apply, although of course the details are somewhat different. Here we consider in detail the more complex of the two cases involving capital accumulation, that in which a renewable resource is used as an input to production. In this case the utilitarian problem takes the form

$$\max \int_0^{\infty} u(c_t, s_t) e^{-\delta t} \, dt$$

s.t. $\dot{k} = F(k_t, \sigma_t) - c_t$ and $\dot{s}_t = r(s) - \sigma_t, \; s_t \geq 0 \, \forall t$

where $F(k_t, \sigma_t)$ is the production function, whose arguments k_t and σ_t are the stock of capital and the flow of the renewable resource used in production. $r(s)$ is the stock growth function for the renewable resource. Then the present value Hamiltonian is

$$H = u(c, s) + \lambda[F(k, \sigma) - c] + \mu[r(s) - \sigma]$$

so that taking account of the first-order conditions on the choice of c and σ, the linearized Hamiltonian and the Hicksian measure of national income is

$$\text{HNI} = \lambda F_k k + \mu r's + u_s s$$

Using the differential equations describing the movement of shadow prices,

$$\dot{\lambda} - \delta \lambda = -\lambda F_k$$
$$\dot{\mu} - \delta \mu = -u_s - \mu r'$$

we have

$$\text{HNI} = k\lambda \left(\delta - \frac{\dot{\lambda}}{\lambda} \right) + s\mu \left(\delta - \frac{\dot{\mu}}{\mu} \right) \tag{11.15}$$

which confirms the general results: national income can be expressed as a return on stocks valued at shadow prices, the rate of return being the difference between the discount rate and the rate of change of the shadow price of the stock, which as noted before is the real rate of return on the stock. As before, the absence of terms in the flows of consumption and the resource do not mean that gifts of these from outside the system would not be valuable. They would of course be valued at rates u_c and $u_c F_\sigma$, and increments of stocks would be valued at rates λ and μ. At the utilitarian stationary state, expression (11.15) reduces to the Fisher-Lindahl-Hicks expression for income as the return on capital (valued at shadow prices):

$$HNI = \delta(k\lambda + s\mu)$$

11.7   Nonutilitarian Objectives

We now have a clear picture of how the Hicksian definition of national income operates in the context of the traditional discounted utilitarian definition of optimality. Given the importance of alternatives to this framework, it is natural to investigate how satisfactory is this definition in the contexts of alternative approaches. Consider first Chichilnisky's definition of intertemporal optimality. We noted in chapter 6 that in the case of an exhaustible resource, the problem

$$\max \alpha \int_0^\infty u(c_t, s_t)e^{-\delta t}\, dt + (1 - \alpha) \lim_{t \to \infty} u(c_t, s_t), \qquad \alpha > 0$$

$$\text{s.t.} \quad \dot{s}_t = -c_t, \qquad s_t \geq 0 \,\forall\, t$$

is equivalent to the utilitarian problem

$$\max \int_0^\infty u(c_t, s_t)e^{-\delta t}\, dt \quad \text{s.t.} \quad \int_0^\infty c_t\, dt \leq \widehat{\beta} s_0$$

initial stock s_0, where $\widehat{\beta}$ is defined in proposition 16.

This means that the optimum according to Chichilnisky's criterion is a utilitarian optimum but for a problem with a different and tighter integral constraint on cumulative consumption. It follows that in this case the results we have for the utilitarian case can be applied to Chichilnisky's criterion, and indeed we have implicitly done this in the previous sections.

In the case of renewable resources, Chichilnisky's criterion, and Hicksian income, matters are more complex. We saw in chapter 7 that a solution exists if and only if the discount rate falls to zero over time. But in the case of a variable discount rate, the equality of the Hamiltonian and the equivalent constant utility level can no longer be established. Hence, we cannot apply the Hicksian income concept in this context. The fact that the discount rate has to go to zero for Chichilnisky's criterion to give solutions, and that as we have noted the Hicksian income concept behaves badly with zero discount rates, gives some intuitive basis for this fact.

In the case of criteria such as the green golden rule and overtaking, the same issues arise: the discount rate is zero, and there is no equality of a Hamiltonian to an equivalent constant utility level.

This leads us to the rather ironic conclusion that the Hicksian income concept, although it is formulated specifically in terms of sustainable utility levels, is not appropriate as an income measure for the optimality criteria that best capture the spirit of sustainability, namely, those placing more weight on the future than the discounted utilitarian criterion.

Chapter 12
National Welfare

In this chapter we turn to the second approach to defining national income, that portrayed in figure 11.1 and associated with the use of the prices defining a separating hyperplane to judge whether a change is an increase in welfare. In the context of an intertemporal economy where consumption bundles are given by functions of time, prices must likewise be functions of time for the entire infinite horizon. In an intertemporal context, this measure is the present value of consumption over time. It is the right-hand side of an intertemporal budget constraint with perfect markets and corresponds more closely to the intuitive concept of wealth than to income. I abbreviate the concept by NW for "national welfare"; we could also think of this as standing for "national wealth." We see that this approach, closer to the standard static interpretation of national income, avoids some of the counterintuitive properties of the Hicksian approach. In particular, its dependence on the discount rate is more natural, and the role of consumption is also more intuitive.

12.1   A Mathematical Framework

We set out the basic principles of measuring national welfare or wealth via the separating hyperplane approach in the context of the model set out in chapter 11, which is repeated here.

Maximize the discounted integral of utilities:

$$\max \int_0^\infty u(c_t, s_t)e^{-\delta t}\, dt$$

subject to the rate-of-change equations

$$\dot{s}_{i,t} = d_i(c_t, s_t), i = 1, \ldots n$$

for the stocks. To solve this problem we constructed a Hamiltonian

$$H = u(c_t, s_t)e^{-\delta t} + \sum_{i=1}^{n} \lambda_{i,t} e^{-\delta t} d_i(c_t, s_t) \tag{12.1}$$

where the $\lambda_{i,t}$ are the shadow prices of the stocks. The first-order conditions for optimality were

$$\frac{\partial u(c_t, s_t)}{\partial c_j} = -\sum_{i=1}^{n} \lambda_{i,t} \frac{\partial d_i(c_t, s_t)}{\partial c_j} \tag{12.2}$$

and

$$\dot{\lambda}_{i,t} - \delta \lambda_{i,t} = -\frac{\partial u(c_t, s_t)}{\partial s_i} - \sum_{k=1}^{n} \lambda_{k,t} \frac{\partial d_k(c_t, s_t)}{\partial s_i} \tag{12.3}$$

The use of argument about separating hyperplanes in problems involving infinite time horizons is mathematically delicate, so we need to be precise about the framework to be used. We assume that the functions $d_i(c_t, s_t), i = 1, \ldots, n$ are such that the set of feasible paths for $c_{j,t}$ and $s_{i,t}$ are bounded. Reasonable conditions sufficient for this are presented for the specific models considered in the discussion of existence of optimal paths in the appendix. Under this assumption, the paths of all variables, including utilities, are such that their integrals against a discount factor with a positive discount rate are finite. Formally, for any i and j,

$$\int_0^\infty c_{j,t} e^{-\delta t}\, dt < \infty, \quad \int_0^\infty s_{i,t} e^{-\delta t} dt < \infty$$

where $c_{j,t}$ and $s_{i,t} : \mathfrak{R} \to \mathfrak{R}$ are real-valued functions of time. We can therefore regard the space of possible paths of consumption levels and stocks as a weighted l_∞ space, with the norm of a function $f(t)$ in this space being given by

$$\|f\| = \sup_t |f(t)e^{-\delta t}|$$

and the inner product of two functions $f(t)$ and $g(t)$ being

$$\langle f, g \rangle = \int_0^\infty f(t)g(t)e^{-\delta t} dt$$

A supporting hyperplane for a set S is then given by a function $h(t)$ such that

$$\langle s(t), h(t) \rangle = \int_0^\infty s(t)h(t)e^{-\delta t} dt \geq 0 \forall s(t) \in S$$

If the function $s(t)$ is a vector-valued function defined on the real numbers, that is, $s(t) : \mathfrak{R} \to \mathfrak{R}^n$, then likewise $h(t) : \mathfrak{R} \to \mathfrak{R}^n$ and $s(t)h(t)$ is interpreted as the inner product of two vectors in $\mathfrak{R}^n$.[1]

A hyperplane that separates the set of paths preferred to an optimum from those that are feasible is a time path of prices for stocks and flows $p_{c,i}(t)$ and $p_{s,i}(t)$ that must satisfy the following conditions (here an asterisk denotes the value of a variable along an optimal path):

$$\int_0^\infty u(c_t, s_t)e^{-\delta t} dt \geq \int_0^\infty u(c_t^*, s_t^*)e^{-\delta t} dt \Rightarrow$$

$$\int_0^\infty [\langle p_c(t), c_t \rangle + \langle p_s(t), s_t \rangle]e^{-\delta t} dt \geq \int_0^\infty [\langle p_c(t), c_t^* \rangle + \langle p_s(t), s_t^* \rangle]e^{-\delta t} dt$$

(12.4)

and

$$[c_t, p_t] \text{ feasible } \Rightarrow$$

$$\int_0^\infty [\langle p_c(t), c_t \rangle + \langle p_s(t), s_t \rangle]e^{-\delta t} dt \leq \int_0^\infty [\langle p_c(t), c_t^* \rangle + \langle p_s(t), s_t^* \rangle]e^{-\delta t} dt$$

(12.5)

where $\langle p_c(t), c_t \rangle$ denotes the inner product of the price vector $p_c(t)$ with the consumption vector c_t. Here condition (12.4) implies that at the prices $p_{c,i}(t)$ and $p_{s,i}(t)$, any path that is at least as good as the optimum path has a present value at least as great as that of the optimum, and

---

[1]Note that in the weighted l_∞ space in which we are working, prices need not always be given by functions of time as implicitly assumed above. Such functions value a consumption path via the inner product, which we have defined. In general, prices are just real-valued linear functions defined on the commodity space. Some of these can be represented by real-valued price functions, as in the text. However, others could be purely finitely additive measures. These are real-valued linear functions defined on the commodity space that value a path as a function only of its limiting properties. For example, $\lim_{t \to \infty} x_t$ is a real-valued linear function on the functions that have limits, and can be extended to those that do not have limits. This point becomes important when we discuss national income in the context of attaining the green golden rule as a policy objective.

condition (12.5) implies that any feasible path has a present value at prices $p_{c,j}(t)$ and $p_{s,i}(t)$ that is no more than that of the optimal path.

DEFINITION 44 A set of prices $p_{c,j}(t)$ and $p_{s,i}(t)$ satisfying (12.4) and (12.5) will be called optimal prices and will be used to define the welfare concept as follows: national welfare along the optimal path is

$$\int_0^\infty \left\{ \langle p_c(t), c_t^* \rangle + \langle p_s(t), s_t^* \rangle \right\} e^{-\delta t} dt$$

By analogy with figure 11.1, we want to establish that any small change that increases this present value welfare measure is a welfare improvement. In proposition 45 we characterize a set of prices that are optimal prices in the sense of the above definition. These prices are intuitive: they are the marginal utilities of the stocks and flows along an optimal path. In effect these marginal utilities define the marginal rates of substitution between the different arguments of the maximand $\int_0^\infty u(c_t, s_t) e^{-\delta t} dt$, and are natural candidates for the role of defining a separating hyperplane.

PROPOSITION 45 The sequence of prices defined by the derivatives of the utility function along an optimal path, that is,

$$\left[p_{c,j}(t), p_{s,i}(t) \right] = \left[\frac{\partial u(c_t^*, s_t^*)}{\partial c_{j,t}}, \frac{\partial u(c_t^*, s_t^*)}{\partial s_{i,t}} \right] \forall j, i, t$$

form a set of optimal prices.

PROOF We need to show that these prices satisfy (12.4) and (12.5). Consider a path (c_t, s_t) such that

$$\int_0^\infty [u(c_t, s_t) - u(c_t^*, s_t^*)] e^{-\delta t} dt \geq 0 \tag{12.6}$$

We need to show that in this case

$$\int_0^\infty \left[\langle p_c(t), c_t \rangle + \langle p_s(t), s_t \rangle \right] e^{-\delta t} dt \geq \int_0^\infty \left[\langle p_c(t), c_t^* \rangle + \langle p_s(t), s_t^* \rangle \right] e^{-\delta t} dt$$

Take a linear approximation to $u(c_t, s_t)$ about (c_t^*, s_t^*) along an optimal path, and use the concavity of the utility function:

$$u(c_t, s_t) - u(c_t^*, s_t^*) \leq \sum_j \frac{\partial u(c_t^*, s_t^*)}{\partial c_{j,t}}(c_{j,t} - c_{j,t}^*) + \sum_i \frac{\partial u(c_t^*, s_t^*)}{\partial s_{i,t}}(s_{i,t} - s_{i,t}^*)$$

$$= \sum_j p_{c,j}(t)(c_{j,t} - c_{j,t}^*) + \sum_i p_{s,i}(t)(s_{i,t} - s_{i,t}^*)$$

Together with (12.6), this establishes the inequality needed, that is, (12.4).

Now we need to establish inequality (12.5). Consider the problem of choosing a program to maximize present value at the prices $[p_{c,j}(t), p_{s,i}(t)]$:

$$\max \int_0^\infty \left[p_c(t)c_t + p_s(t)s_t \right] e^{-\delta t} \, dt$$

$$\dot{s}_{i,t} = d_i(c_t, s_t), i = 1, \ldots, n$$

The corresponding Hamiltonian is

$$H = \left[\langle p_c(t), c_t \rangle + \langle p_s(t), s_t \rangle \right] e^{-\delta t} + \sum_{i=1}^n \mu_{i,t} e^{-\delta t} \, d_i(c_t, s_t)$$

and a solution satisfies

$$p_{c,j}(t) = -\sum_{i=1}^n \mu_{i,t} \frac{\partial d_i(c_t, s_t)}{\partial c_j}$$

and

$$\dot{\mu}_{i,t} - \delta \mu_{i,t} = -p_{s,i}(t) - \sum_{k=1}^n \mu_{k,t} \frac{\partial d_k(c_t, s_t)}{\partial s_i}$$

Now note that given the definition of the optimal prices, these conditions are precisely the same as conditions (12.2) and (12.3), which characterize a solution to the general optimization problem of maximizing (11.2) subject to (11.1). Hence, a path that solves the overall optimization problem also solves the problem of maximizing the present value of the path at the optimal prices. This completes the proof.

We have now established that the derivatives of the utility function with respect to stocks and flows can be used to define a hyperplane that separates the set of paths preferred to an optimal path from the set of feasible paths. They can therefore be used to define a price system at which national welfare can be computed. It is of course immediate that any small change in a path that has a positive present value at these optimal prices increases national welfare.

COROLLARY 46 Let a small variation $(\Delta c_t, \Delta s_t)$ about an optimal path (c_t^*, s_t^*) have positive present value at the optimal prices $[p_{c,j}(t), p_{s,i}(t)]$. Then the implementation of this variation leads to an increase in welfare. Conversely, if a small variation $(\Delta c_t, \Delta s_t)$ about an optimal path (c_t^*, s_t^*) leads to an increase in welfare, then it has positive value at the optimal prices $[p_{c,j}(t), p_{s,i}(t)]$.

PROOF By assumption

$$\int_0^\infty [\langle \Delta c_t, p_c(t) \rangle + \langle \Delta s_t, p_s(t) \rangle] e^{-\delta t}\, dt > 0 \qquad (12.7)$$

Let $(\Delta c_t + c_t^*, \Delta s_t + s_t^*)$ be the path resulting from implementing the variation in the optimal path. The welfare associated with this path is

$$\int_0^\infty [u(c_t^*, s_t^*) + \sum_j \frac{\partial u(c_t^*, s_t^*)}{\partial c_{j,t}} \Delta c_{j,t} + \sum_i \frac{\partial u(c_t^*, s_t^*)}{\partial s_{i,t}} \Delta s_{i,t}] e^{-\delta t}\, dt$$

which by (12.7) and the definition of the optimal prices is greater than that on the optimal path, as required. The proof of the converse is immediate.

12.2 National Welfare and Resources

It is natural to inquire in more detail about the relationship between the two definitions and about the connection, if any, that national welfare has with the Hamiltonian, which played so central a role in the definition of Hicksian income. This is best done by comparing the two definitions in the context of specific models, so we begin with a brief analysis of the implications of our definition of national welfare for resource-based economies, using the same frameworks as we used for the Hicksian definition of national income.

12.2.1 National Welfare and Hicksian Income in the Hotelling Case
— As an exercise whose results are quite interesting and neat, we consider briefly the present value of national welfare, in the sense defined above, in the case of the classic formulation due to Hotelling, summarized in chapter 2. In this case, we seek to

$$\max \int_0^\infty u(c_t)e^{-\delta t}\,dt \quad \text{s.t.} \quad \int_0^\infty c_t\,dt = s_0$$

Recall that this is a model in which the remaining stock of the resource is not supposed to be a source of benefits to the economy. Letting the optimal path of consumption be denoted by an asterisk, the present value of national welfare would in this case be measured by

$$\text{PVNW} = \int_0^\infty c_t^* u'(c_t^*)e^{-\delta t}\,dt$$

Noting that $u'(c_t^*)e^{-\delta t}$ is a constant, equal to the initial value of the shadow price p_0, this is simply the initial stock of the resource multiplied by the initial shadow price:

$$\text{PVNW} = p_0 s_0$$

This is an extremely simple and natural result: the welfare the economy can attain depends on its stock and the social value of this stock. This is a very reasonable measure of its wealth.

What is the present value of the Hicksian measure of national income in this case? The Hamiltonian is

$$H = u(c_t)e^{-\delta t} - p_t c_t$$

so that the linearized Hamiltonian is zero. Hence, the present value of Hicksian income is zero also.

So in this case the Hamiltonian-based Hicksian measure of present value national income is zero, and the welfare or wealth measure is the shadow value of the initial stock of the resource.

This comparison shows clearly that there are two distinct concepts here: national welfare and Hicksian income. In the Hotelling case the latter is zero whereas the former is positive and depends on the stock and the utility function (which determines the shadow price).

Note that although there is a radical difference in these measures, the shadow prices associated with these two approaches are the same: in each case the shadow price of the single good is the derivative of the utility function with respect to its consumption along an optimal path. However, Hicksian national income remains zero even if the initial stock of the resource is doubled, whereas the national welfare or wealth measure increases.[2] But the identity of the shadow prices in the two cases is reassuring. It means that if we use the value of a change at shadow prices to judge whether it is an improvement, we will arrive at the same conclusion in both cases.

12.2.2 Exhaustible Resources and National Welfare — Consider first an economy with only exhaustible resources as a source of consumption, which as in chapter 3 values these both as a stock and as a flow. Its optimal path in the utilitarian sense is defined by the problem

$$\max \int_0^\infty \left[u_1(c_t) + u_2(s_t) \right] e^{-\delta t} \, dt$$

subject to

$$\dot{s}_t = -c_t$$

Let $\left(c_t^*, s_t^* \right)$ be a solution to this problem; then the national welfare along this path is

$$\text{NW} = \int_0^\infty [c_t^* u_1'(c_t^*) + s_t^* u_2'(s_t^*)] e^{-\delta t} \, dt$$

This is automatically a present value because the separating hyperplane defines prices covering the entire infinite horizon.

How does this expression relate to the Hicksian interpretation of national income? The Hicksian interpretation of national income is

---

[2]But does not double: in the formula PVNW $= p_0 s_0$, doubling the stock generally changes, indeed reduces, the initial price.

an instantaneous measure of income, holding at a particular time t. Obviously the instantaneous income level corresponding to NW is $[c_t^* u_1'(c_t^*) + s_t^* u_2'(s_t^*)]$. Note that on an optimal path the shadow price of the resource that features in the Hamiltonian, denoted λ_t in the previous sections and chapters, equals the marginal utility of consumption: $\lambda_t = u_1'(c_t^*)$. Furthermore, the marginal utility of the stock can be expressed as a function of λ_t and its rate of change: $u_2'(s_t^*) = \delta\lambda_t - \dot{\lambda}_t$. Hence the instantaneous value of NW is

$$\text{NW}_t = c_t^* \lambda_t + s_t^* \lambda_t \left(\delta - \frac{\dot{\lambda}_t}{\lambda_t}\right)$$

This is precisely the Hicksian national income HNI *plus* the term $c_t^* \lambda_t$; it is the flow of utility from the stock plus the value of consumption. Of course, in a stationary state, the two are equal, as $c_t = 0$. Then we have

$$\text{NW}_t = s_t^* \lambda_t \delta$$

the classic Hicksian formulation but now being interpreted as a coordinate of a separating hyperplane and a measure of welfare. In general, however, the relationship between the two cases is not so close.

12.2.3 Renewable Resources and National Welfare — Consider next the by-now familiar problem

$$\max \int_0^\infty u(c_t, s_t) e^{-\delta t} \quad \text{s.t.} \quad \dot{s}_t = r(s_t) - c_t$$

On the above definition, national welfare is measured by

$$\text{NW} = \int_0^\infty [u_c(c_t^*, s_t^*) c_t + u_s(c_t^*, s_t^*) s_t] e^{-\delta t} \, dt$$

How does this compare with the Hicksian definition? For the same problem, the Hicksian measure is

$$\text{HNI}_t = s_t \lambda_t \left(\delta - \frac{\dot{\lambda}_t}{\lambda_t}\right)$$

where of course λ_t is the shadow price of the resource accumulation constraint. The instantaneous national welfare measure NW_t can be expressed as

$$NW_t = c_t \lambda_t + s_t \lambda_t (\delta - r' - \frac{\dot{\lambda}_t}{\lambda_t})$$

At a stationary solution with a positive discount rate, $HNI_t = s_t \lambda_t \delta$ and $NW_t = c_t \lambda_t + s_t \lambda_t (\delta - r')$. At a stationary solution when the discount rate falls asymptotically to zero, the stationary solution is the green golden rule, so that $\dot{\lambda}_t = \delta = 0$, implying that $NW_t = c_t \lambda_t - \lambda_t s_t r'(s_t)$ and $HNI_t = 0$. In fact the difference between NW_t and HNI_t is always $c_t \lambda_t - \lambda_t s_t r'(s_t)$, which is the flow of consumption minus the return on the stock, evaluated at the marginal productivity of the stock in generating consumption flows. Recall that to the right of the maximum sustainable yield, $r'(s) < 0$, so that the difference is always positive in this case. Clearly these two expressions are quite different, and are measuring different characteristics of the economy. As I noted before, the national welfare measure does not depend on the discount rate in the way that the Hicksian measure does.

12.2.4 National Welfare and Capital Accumulation — Finally we consider the case that generalizes the previous one to include production and accumulation of capital. Formally, the economy's aim is to

$$\max \int_0^\infty u(c_t, s_t) e^{-\delta t} \, dt$$

s.t. $\dot{k}_t = F(k_t, \sigma_t) - c_t$ and $\dot{s}_t = r(s_t) - \sigma_t, s_r \geq 0 \, \forall t$

In this case national welfare is again

$$NW = \int_0^\infty \left[u_c(c_t^*, s_t^*) c_t + u_s(c_t^*, s_t^*) s_t \right] e^{-\delta t} \, dt$$

whereas from (11.15) Hicksian national income is

$$HNI_t = k_t \lambda_t \left(\delta - \frac{\dot{\lambda}_t}{\lambda_t} \right) + s_t \mu_t \left(\delta - \frac{\dot{\mu}_t}{\mu_t} \right)$$

where λ_t is the shadow price on the resource and μ_t that on the capital stock. Instantaneous national welfare NW_t can be expressed as

$$\mathrm{NW}_t = c_t \lambda_t + s_t \mu_t \left(\delta - \frac{\dot{\mu}}{\mu} - r' \right)$$

As in the previous section, national welfare includes the terms $c_t \lambda_t$ and $s_t \mu_t r'$, which have no analogue in the expression for Hicksian national income; however, the latter includes a term in the return on the capital stock, and the former is independent of the capital stock. Once again, the Hicksian national income is zero at a stationary solution at which the discount rate is zero.

12.3   Chichilnisky's Criterion

The discussion of national income concepts so far has been within the framework of discounted utilitarianism. Now it is time to extend it to more general cases. Here we focus on the case of an economy that defines optimality according to Chichilnisky's criterion, involving both an integral of utilities and a term depending on long-run utility. The next section addresses the same issues in the context of an economy seeking to attain the maximum sustainable utility level.

Recall that with Chichilnisky's definition of optimality, two distinct cases emerge: the case of exhaustible resources and the case of renewable resources. In the latter case, we noted that an optimum exists only if the discount rate in the integral of utilities goes to zero in the limit, in which case the maximum value of the term in limiting utility is automatically attained by the path that is a utilitarian optimum. In this case, all the analysis of national income concepts of the previous sections applies without alteration.

More interesting and challenging is the case of exhaustible resources. Consider the simplest version of this case:

$$\max \int_0^\infty u(c_t, s_t) e^{-\delta t} \, dt + \lim_{t \to \infty} u(c_t, s_t) \quad \text{s.t.} \quad \dot{s}_t = -c_t \text{ and } s_t \geq 0 \,\forall t$$

Let (c_t^*, s_t^*) be the optimal path for this problem. National welfare is now defined as

$$\mathrm{NW} = \int_0^\infty \left[c_t u_c(c_t^*, s_t^*) + s_t^* u_s(c_t^*, s_t^*) \right] e^{-\delta t} \, dt$$
$$+ \lim_{t \to \infty} \left[c_t u_c(c_t^*, s_t^*) + s_t u_s(c_t^*, s_t^*) \right]$$

$$(12.8)$$

The prices that are used to define national welfare contain two elements: one corresponding to the integral term, as in the case of the discounted utilitarian approach, and an extra element arising from the value placed by the objective on the limiting utility level. This second element is related to a point made in footnote 1 on p. 177, namely that in the space we are using, prices may not always be given by functions of time that define present values; they can instead be given by purely finitely additive measures, linear functions that assign to a path of consumption or stock or utility a value that depends only on its limiting properties. In fact the objective $\max \lim_{t \to \infty} u(c_t, s_t)$ is similar, as it evaluates a utility sequence only by reference to its limiting value. Prices may also be given by mixtures of the two types of function, which is just what we have in the measure of national welfare in (12.8). Here the value assigned by prices to a path of the economy depends both on the time path over finite horizons, via the present value term, and on the limiting values along the path. Formally, we are now defining the value of a sequence of consumption and stock levels (c_t, s_t) at prices $p_c(t), p_s(t)$, or equivalently defining the inner product of the consumption and stock sequences $\langle (c_t, s_t), [p_c(t), p_s(t)] \rangle$ with the prices sequences, as

$$\langle (c_t, s_t), [p_c(t), p_s(t)] \rangle$$

$$= \int_0^\infty [c_t p_c(t) + s_t p_s(t)] e^{-\delta t} dt + \lim_{t \to \infty} [c_t p_c(t) + s_t p_s(t)] \quad (12.9)$$

We then define a supporting hyperplane for a set S of paths (c_t, s_t) of consumption and of the resource stock as functions $p_c(t), p_s(t)$ such that

$$\int_0^\infty [c_t p_c(t) + s_t p_s(t)] e^{-\delta t} dt + \lim_{t \to \infty} [c_t p_c(t) + s_t p_s(t)] \geq 0 \, \forall \, (c_t, s_t) \in S$$

With Chichilnisky's definition of optimality, the price system contains undiscounted terms because of the limiting term in the definition of optimality. So national welfare is measured in (12.8) as a present value, plus a term reflecting long-run or sustainable welfare. This term is not discounted, although apart from this it has the same form as the other terms, namely, flows evaluated at prices given by marginal valuations along an optimal path. The presence of this extra term is important because it gives a reason for using in the measurement of national welfare prices that relate to the distant future but are not discounted.

All the statements in this section can readily be formalized using the mathematical framework set up in section 12.1. Recall that there we gave formal definitions of separating hyperplanes and optimal prices, and then proved that the time path of marginal utilities forms a separating hyperplane and an optimal price system, defining national welfare. We can paraphrase the definitions and arguments there, which we do here briefly.

A hyperplane that separates the set of paths preferred to an optimum from those that are feasible is a time path of prices for stocks and flows $p_{c,j}(t)$ and $p_{s,i}(t)$, which must satisfy the following conditions (here an asterisk denotes the value of a variable along an optimal path):

$$\int_0^\infty u(c_t, s_t)e^{-\delta t}\, dt + \lim_{t \to \infty} u(c_t, s_t)$$

$$\geq \int_0^\infty u(c_t^*, s_t^*)e^{-\delta t}\, dt + \lim_{t \to \infty} u(c_t^*, s_t^*)$$

$$\Rightarrow \langle (c_t, s_t), [p_c(t), p_s(t)] \rangle \geq \langle (c_t^*, s_t^*), [p_c(t), p_s(t)] \rangle$$

i.e., $$\int_0^\infty [p_c(t)c_t + p_s(t)s_t]e^{-\delta t}\, dt + \lim_{t \to \infty} [p_c(t)c_t + p_s(t)s_t]$$

$$\geq \int_0^\infty [p_c(t)c_t^* + p_s(t)s_t^*]e^{-\delta t}\, dt + \lim_{t \to \infty} [p_c(t)c_t^* + p_s(t)s_t^*]$$

$$(12.10)$$

and

$$(c_t, p_t) \text{ feasible implies}$$

$$\langle (c_t, s_t), (p_c(t), p_s(t)] \rangle \leq \langle (c_t^*, s_t^*), [p_c(t), p_s(t)] \rangle \qquad (12.11)$$

where $\langle p_c(t), c_t \rangle$ denotes the inner product (12.9) of the price vector $p_c(t)$ with the consumption vector c_t. Here condition (12.10) implies that at prices $p_{c,j}(t)$ and $p_{s,i}(t)$, any path that is at least as good as the optimum path has a value (present value plus limiting value) at least as great as that of the optimum, and condition (12.11) implies that any feasible path has a value (present value plus limiting value) at prices $p_{c,j}(t)$ and $p_{s,i}(t)$ that is no more than that of the optimal path.

DEFINITION 47 A set of prices $p_{c,j}(t)$ and $p_{s,i}(t)$ satisfying (12.10) and (12.11) will be called optimal prices and will be used to define national welfare as follows: national welfare along the optimal path is the inner product (12.9) of the optimal prices with the optimal paths of consumption and stocks:

$$NW = \langle (c_t^*, s_t^*), [p_c(t), p_s(t)] \rangle$$

$$= \int_0^\infty [p_c(t)c_t^* + p_s(t)s_t^*]e^{-\delta t}\, dt + \lim_{t \to \infty}[p_c(t)c_t^* + p_s(t)s_t^*]$$

By analogy with figure 11.1, we now want to establish that any small change that increases this welfare measure is a welfare improvement. In proposition 48 we show that the marginal utilities of the stocks and flows along an optimal path are optimal prices in the sense of definition 47. The fact that a small change that leads to an increase in this national income measure is a welfare improvement is then immediate.

PROPOSITION 48 The sequence of prices defined by the derivatives of the utility function along an optimal path, that is,

$$[p_{c,j}(t), p_{s,i}(t)] = \left(\frac{\partial u(c_t^*, s_t^*)}{\partial c_{j,t}}, \frac{\partial u(c_t^*, s_t^*)}{\partial s_{i,t}} \right) \forall\, j, i, t$$

form a set of optimal prices in the sense of definition 47.

PROOF We need to show that these prices satisfy (12.10) and (12.11). Consider a path (c_t, s_t) such that

$$\int_0^\infty [u(c_t, s_t) - u(c_t^*, s_t^*)]e^{-\delta t}\, dt + \lim_{t \to \infty}[u(c_t, s_t) - u(c_t^*, s_t^*)] \geq 0 \quad (12.12)$$

We need to show that in this case

$$\int_0^\infty [p_c(t)c_t + p_s(t)s_t]e^{-\delta t}\, dt + \lim_{t \to \infty}[p_c(t)c_t + p_s(t)s_t]$$

$$\geq \int_0^\infty [p_c(t)c_t^* + p_s(t)s_t^*]e^{-\delta t}\, dt + \lim_{t \to \infty}[p_c(t)c_t^* + p_s(t)s_t^*]$$

Take a linear approximation to $u(c_t, s_t)$ about (c_t^*, s_t^*) along an optimal path, and use the concavity of the utility function:

$$u(c_t, s_t) - u(c_t^*, s_t^*) \leq \sum_j \frac{\partial u(c_t^*, s_t^*)}{\partial c_{j,t}}(c_{j,t} - c_{j,t}^*) + \sum_i \frac{\partial u(c_t^*, s_t^*)}{\partial s_{i,t}}(s_{i,t} - s_{i,t}^*)$$

$$= \sum_j p_{c,j}(t)(c_{j,t} - c_{j,t}^*) + \sum_i p_{s,i}(t)(s_{i,t} - s_{i,t}^*)$$

Together with (12.12), this establishes the inequality needed, that is, (12.10).

Inequality (12.11) can now be established using a very minor variation of the argument used for proposition 45 of section 12.1.

It is now immediate that any small change in a path that has a positive present value at these optimal prices increases national welfare:

COROLLARY 49 Let a small variation $(\Delta c_t, \Delta s_t)$ about an optimal path (c_t^*, s_t^*) have positive inner product with (i.e., positive value at) the optimal prices $[p_{c,j}(t), p_{s,i}(t)]$. Then the implementation of this variation leads to an increase in welfare.

PROOF The proof is immediate.

In summary, the definition of national income implied by Chichilnisky's criterion of intertemporal optimality involves the use of prices to assign value to a path of the economy. The value has two components, one a present value computed at a discount rate in the conventional fashion, and one an undiscounted value associated with the very long-run properties of the path.

12.4 Sustainable Revenues and National Income

Suppose now that the economy is very future oriented, in that the objective is to achieve the maximum sustainable utility, the solution to which is, as we have seen, to move to the green golden rule. What prices, and what private behavior by agents in the economy, will support such an outcome? And how do we measure national income in this case? The outcome in this case is an extension of that of section 12.3, in which we considered national welfare in an economy following Chichilnisky's approach to intertemporal optimality. Now welfare and the optimal prices depend only on the long-run or limiting terms.

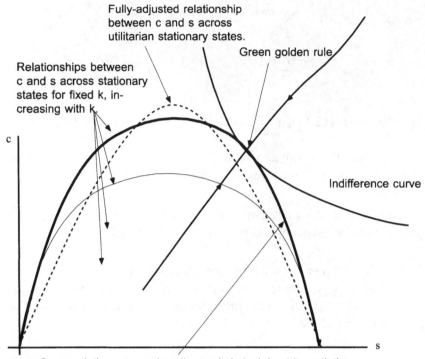

Fully-adjusted relationship
between c and s across
utilitarian stationary states.

Green golden rule

Relationships between
c and s across stationary
states for fixed k, in-
creasing with k.

Indifference curve

c

s

Curve relating c to s when the capital stock is at its satiation
level: this is the upper boundary of the c-s curves for fixed
values of the capital stock.

FIGURE 12.1   The optimal path according to the Chichilnisky criterion and the
utilitarian criterion with the discount rate going to zero.

Before a formal analysis in mathematical terms, consider for a moment
figure 10.2, repeated here as figure 12.1. The green golden rule appears
as a point of tangency between a convex feasible set (the set of points on
or below the graph of the growth function) and a convex preferred-or-
indifferent set.

From standard microeconomic theory, we expect to be able to support
such a point as in figure 12.1 by facing agents with relative prices for
the two commodities (the stock and the flow in this case) equal to the
common slope of both sets at their point of tangency. Applying this in
the present context, we could normalize the price of consumption to be
unity, and set the price of the stock to be $p = r'(s^*)$. Maximizing

$$c + ps, \text{ where } p = r'(s^*)$$ (12.13)

over the set of *c–s* points feasible in the long run will lead to the green golden rule. In the renewable resource context, think of the following example. The owner of a resource stock can sell a flow generated from this, and is also paid a rent for maintaining the stock. The stock might be a forest. The flow would be derived by cutting and selling a part of this, and the rental payment on the stock could be payments made by people using the forest for recreational purposes, a payment in recognition of the forest's carbon sequestration role, or a preservation payment made to the owner by the government. Then if the rent relative to the price of the flow is $r'(s^*)$, the combination of stocks and flows that maximizes total receipts is the green golden rule (see figure 12.2).

This seems very clear; in fact, the clarity is deceptive. There is a difficult point here, relating to the time dimension. The diagram in figure 12.1, and the analysis here, relate to a one-period framework. However, we

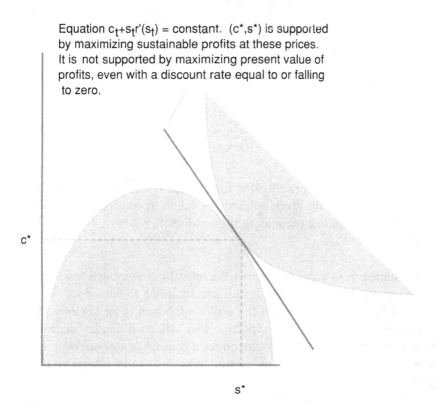

Equation $c_t + s_t r'(s_t) =$ constant. (c^*, s^*) is supported by maximizing sustainable profits at these prices. It is not supported by maximizing present value of profits, even with a discount rate equal to or falling to zero.

c^*

s^*

FIGURE 12.2   The green golden rule maximizes sustainable utility, and max-imizes one-period value at the prices defined by the separating line.

are interested in behavior that supports the green golden rule forever. We would like to say that choosing (c^*, s^*) maximizes the sum of revenues from the resource over the long run, but we cannot say this, as this sum is clearly infinite, and there are many other feasible (c, s) combinations that will also give an infinite value. And we cannot say that (c^*, s^*) maximizes the present discounted value of revenues from the resource because it does not: for any positive discount rate, the policy that maximizes the present value of profits is nonstationary.

What can we say? We can say that the green golden rule leads to the highest indefinitely maintainable level of revenues from the use of the resource: it maximizes "sustainable" or limiting or long-run revenues.

There is an important conclusion here: if society is so future oriented as to wish to support the highest sustainable utility level (i.e., the green golden rule), then we need correspondingly future oriented behavior on the parts of agents in the economy. We need firms to seek the highest sustainable profits (i.e., the maximum value of profits that can be maintained forever) and resource owners to manage their resources so as to yield the highest sustainable revenues from the resources. Formally,

PROPOSITION 50 Consider the economy described by

$$\max \lim_{t \to \infty} u(c_t, s_t)$$

$$\text{s.t. } \dot{s}_t = r(s) - c_t, s_t \geq 0 \, \forall t$$

Then there exists a price for the flow of the resource and a rental for the stock such that the green golden rule values of consumption and the resource stock lead to the maximum sustainable revenues from the use of the resource.

In summary, in order to support the most future-oriented framework in the context of renewable resources, we need to introduce a new form of behavior for agents in the economy, which in an obvious sense corresponds to the social perspective and reflects a similar degree of future orientation. We need to introduce a concept of sustainable profits and revenues, and assume that agents behave so as to maximize these goals.

How do these observations relate to the previous discussions of national income in its several interpretations? As we have noted, with a zero discount rate, the Hicksian national income is not well defined. It is not applicable here, although this seems precisely the type of case in which it would be appropriate because we are considering an economy oriented to maximizing the sustainable welfare level.

What we are using here is again the separating hyperplane approach, as discussed in the previous section addressing national welfare and Chichilnisky's criterion, but now instead of the price system consisting of two parts, one defining a present value via an integral and the other reflecting the limiting behavior of the path, we now have only the latter term. The national welfare measure corresponding to this price system is now

$$\lim_{t \to \infty} [c_t u_c(c_t^*, s_t^*) + s_t u_s(c_t^*, s_t^*)] \tag{12.14}$$

which can be rewritten as

$$\lim_{t \to \infty} [c_t + p_t s_t], \text{ where } p = r'(s^*) = \frac{u_s(c_t^*, s_t^*)}{u_c(c_t^*, s_t^*)}$$

This is precisely the price system introduced in (12.13), which we can now see as a particular form of our earlier concept of national welfare.

12.5 Summary

The following important points have emerged in the discussion of national income in this and the preceding chapter:

- There are two alternative concepts: the Hicksian measure of national income, which is related to the Hamiltonian of a dynamic optimization problem, and the national welfare concept, which is the generalization to dynamic economies of the welfare concept based on separating hyperplanes and used in welfare economics. It is a measure of a society's wealth rather than income.*

---

*Since completion of this manuscript the relationship between Hicksian income (HNI) and national welfare (NW) has been characterized. Heal and Kristrom have shown that if a policy change at date zero leads to changes ΔHNI and ΔNW in HNI and NW, respectively, then

$$\Delta \text{HNI} = \delta \text{NW}$$

Changes in the Hamiltonian-based income are therefore interest on changes in the hyperplane-based wealth. We also show that ΔNW equals the integral of the present values of changes in HNI along an optimal path minus the integral of the present value of changes in stock accumulation along the path.

- The appropriate measure depends on the objective of the economy (utilitarian, Rawlsian, Chichilnisky), and in the cases of utilitarian and Chichilnisky formulations it of course depends on the discount rate. Different objectives and different discount rates give different numbers, or even different formulae, for national income. Hicksian income can be measured in terms of the stocks in an economy; however, national welfare requires data on consumption. The pure depletion model of Hotelling illustrates nicely the distinction: Hicksian or sustainable income is zero, whereas national welfare or wealth depends on the initial stock of the resource. However, the shadow prices of goods and services are the same in both approaches.
- If we adopt Chichilnisky's criterion, then the present value national welfare measure of national income involves the use of undiscounted future values. This approach provides a justification for not discounting some of the services provided in the future by environmental assets.
- An increase in national income as measured by national welfare NW indicates an increase in welfare according to the corresponding objective function for small changes in the values of variables; technically, it is a first-order approximation of the true measure of welfare.
- The increase in welfare is Paretian in the sense that there is a net gain in welfare: the gainers could compensate the losers and remain gainers.
- The Hicksian measure of national income can be expressed entirely in terms of stocks—of physical capital and of natural capital. It reduces to the Fisher–Lindahl–Hicks measure of a return on the value of a stock.
- When the economy's objective is the maximization of long-run welfare (i.e., obtaining the green golden rule), then the optimal configuration cannot be supported by the maximization of present value. We need to introduce the concept of maximum sustainable profits, and there exist prices at which the maximization of sustainable profits corresponds to and supports the maximization of sustainable utility.

Several points of practical significance are implied by these observations. Stocks are a sufficient statistic for the measurement of Hicksian national income. This emphasizes the importance of accurate measurement of all aspects of an economy's stocks, including its stocks of environmental assets. Some of these are global public goods, such as the atmosphere. This is a task that we have barely begun; it is probably one of the biggest challenges facing environmental scientists. In this context, it is worth referring to a very enterprising recent World Bank study [100] that seeks in a preliminary way to evaluate the contributions of physical capital, natural capital, and human capital to the wealth of nations.

Within recent years there have been moves to recalculate national income on a "green" basis for several countries. For example, a study by Repetto et al. [93] for Costa Rica recently argued that when environmental costs and benefits are fully recognized, Costa Rica's national income is reduced significantly. There has been a similar study for Indonesia [94]. These studies suggest that the issues analyzed here are important in practice. In fact, the studies probably underestimate their importance, as data limitations allow them to take account of changes in only a small number of environmental assets.

Finally, a caution is in order on the use of national income measures for evaluating the consequences of far-ranging phenomena such as climate change. National income is a local measure of welfare changes, a first-order approximation. It seems to be quite within the bounds of scientific possibility that within 150 to 200 years, climate change could make nonlocal changes to our economic environment. For example, a recent study by Chichilnisky et al. [24] suggested that without the implementation of countervailing policies, exposure to inundation could change radically over several centuries, and over a shorter period agricultural productivity might drop by as much as one third in India. National income as computed in this chapter is not an appropriate measure of the impact of such changes; the full impacts can be captured only by a general equilibrium model with a complete description of consumer demands.

The issue of how to evaluate the impact of policy changes leads naturally to the matter of cost–benefit analysis or project evaluation, to which we turn in chapter 13.

Chapter 13
Project Evaluation

The essence of project evaluation is deciding which of a set of alternative projects to undertake. Naturally, we wish to undertake those that will contribute most to national welfare. A key step is therefore measuring this contribution. For this we use shadow prices. As we noted in chapter 12, corollary 46, a necessary condition for a small change to increase national welfare is that its value be positive at the shadow prices used to measure national welfare.

Smallness is critical here: shadow prices are first-order or linear approximations to relationships that may be highly nonlinear. This is clear from proposition 45, which defines the shadow prices for the national welfare measure as the derivatives of the utility function evaluated along an optimal path. Consequently, the results of far-reaching changes such as damage to important life-support ecosystems (as may be caused by global warming) may be too extensive to be valued at shadow prices. As the benefits from projects designed to prevent such changes are just the avoidance of these far-reaching changes, the same applies to these projects. To evaluate such projects, we need more than linear approximations: we need models with complete functional relationships between the key variables.

There is a potential ambiguity in this approach to project evaluation because we have identified two possible definitions of national income: the Hicksian income measure and the national welfare level. However, as we have noted, in the case of utilitarian objectives the shadow prices are the same for both approaches. In the case of more future-oriented criteria such as Chichilnisky's, the green golden rule, and the overtaking criterion, then the Hicksian definition seems inappropriate. So to avoid uncertainty, we assume that we are working with shadow prices derived from the national welfare measure of chapter 12; this provides a framework that covers all possible objectives and anyway links more closely with the traditional concerns of welfare economics.

A rigorous framework for project evaluation or cost–benefit analysis should clearly be based on an analysis of optimal natural resource use patterns and involves using models of the types in the earlier chapters to provide

- A rigorous analysis of the relationships between the shadow prices of different commodities
- A statement of how these should move over time
- An analysis of how shadow prices relate to the objective chosen by society, such as the nature of the objective function and the discount rate
- Analysis of how shadow prices depend on physical constraints such as production functions and resource supply conditions.

In view of this, in this chapter we focus on what our earlier analysis has told us about shadow prices and their determinants and dynamics. We focus particularly on the impact on shadow prices of two features that we have identified as crucial to the analysis of sustainability: recognizing the value of environmental assets and valuing the long-run.

13.1 Exhaustible Resources

The simplest framework for exploring this issue is given by the set of models of depletion of an exhaustible resource analyzed in chapters 2 and 3. For these models, four cases in total were reviewed in chapters 2, 3, and 6:

- Depletion of an exhaustible resource with only the rate of use a source of utility, the original Hotelling case
- Depletion of an exhaustible resource with the flow and the stock as sources of utility
- Depletion with Chichilnisky's criterion
- Depletion aimed at the green golden rule, which selects the highest sustainable utility level

The Rawlsian and overtaking definitions of optimality were also considered, and led to the same policies as the green golden rule. The policies optimal in the four cases itemized here are compared in figure 13.1. Three of the four paths in figure 13.1—the discounted utilitarian

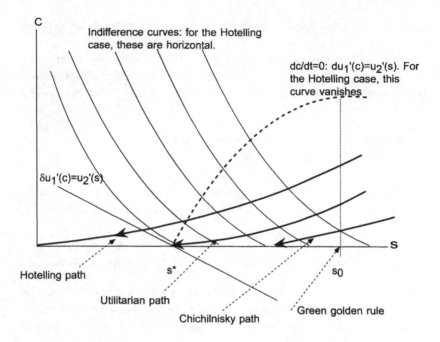

FIGURE 13.1 Optimal depletion paths for alternative concepts of optimality, ranging from Hotelling's formulation to the green golden rule.

the Chichilnisky, and the green golden rules cases—are in figure 6.3 of chapter 6. The addition is the Hotelling case. It is easy to accommodate this case in the diagram: the marginal utility of the stock is identically zero as the stock is not valued, that is, $u_2'(s) \equiv 0$, so that the set of points satisfying $\delta u_1' = u_2'$, the set on which $\dot{c}$ changes sign from negative to positive according to the differential equations governing the optimal consumption paths, is empty, and consequently the consumption level falls at all points in the c–s plane. Both the consumption level and the remaining stock go to zero in the limit in this case.

13.1.1 Initial Shadow Prices — Note that the initial consumption level on an optimal path depends on the optimality criterion. It is greatest on the Hotelling path, smaller on the path corresponding to valuation of the stock with the discounted utilitarian criterion, smaller again on the path optimal according to the Chichilnisky criterion, and zero according to the green golden rule. Now, from the condition for the maximization of the Hamiltonian with respect to consumption,

$$u_1'(c_0) \leq \lambda_0, \ = \ 0 \text{ if } c_0 > 0$$

that is, the initial shadow price is greater than or equal to the initial marginal utility of consumption, with equality if the initial consumption level is positive. It follows, then, that shadow prices are ranked in the opposite order to the initial consumption level in these four cases: the initial shadow price is highest in the case of the maximization of long-run utility, less high with Chichilnisky's criterion, lower again in the case of maximization of a discounted sum of utilities with the stock a source of utility, and lowest of all in the Hotelling case. In addition, we can see that within each of the last three cases, all of which involve a discount rate, the initial shadow price depends on the discount rate. In these cases, the higher the discount rate, the lower the initial shadow price of the resource. For the Hotelling problem (chapter 2) and the discounted utilitarian case with resource stock an argument of the utility function (chapter 3), the proof of this is immediate from the arguments in the earlier chapters. For the case of Chichilnisky's criterion, it follows from the characterization of a stationary resource stock (6.4), which is repeated here:

$$u_2'(\hat{s}) = u_1'(0)\frac{\alpha\delta}{(1 - \alpha)\delta + \alpha}$$

Differentiating this with respect to the discount rate, we find that

$$\frac{\partial u_2'(\hat{s})}{\partial \delta} = \frac{\alpha^2}{[(1 - \alpha)\delta + \alpha]^2} > 0$$

so that the steady-state resource stock is decreasing in the discount rate δ. From this it follows that the initial consumption level is higher, and the initial shadow price lower, the greater is the discount rate. This is summarized in the following proposition.

PROPOSITION 51 An economy with a fixed stock s_0 of an exhaustible resource uses this over time so as to maximize an objective function, which may be either

$$\lim_{t \to \infty} u(c_t, s_t) \text{ (green golden rule, GGR), or}$$

$$\int_0^\infty u(c_t, s_t) e^{-\delta t} dt + \lim_{t \to \infty} u(c_t, s_t) \text{ (Chichilnisky, Ch), or}$$

$$\int_0^\infty u(c_t, s_t) e^{-\delta t} dt \text{ (discounted utilitarian, DU), or}$$

$$\int_0^\infty u(c_t) e^{-\delta t} dt \text{(Hotelling, H)}$$

The constraint is $\int_0^\infty c_t \, dt \leq s_0$. Then the shadow prices λ_0 of the resource at the initial date corresponding to the alternative maximization criteria satisfy the following inequalities:

$$\lambda_0(\text{GGR}) > \lambda_0(\text{Ch}) > \lambda_0(\text{DU}) > \lambda_0(\text{H})$$

In each of the three cases in which there is a discount rate, the initial shadow price λ_0 is lower, the higher is the discount rate.

Table 13.1 summarizes these conclusions.

Table 13.1

Case	Values Short Run	Values Long Run	Values Assets	Initial Price
1. Hotelling	Yes	No	No	Lowest
2. Stock valued	Yes	No	Yes	High
3. Chichilnisky	Yes	Yes	Yes	Higher
4. GGR	No	Yes	Yes	Highest

So *the effect of emphasizing considerations relevant to sustainability is to raise the shadow price of the exhaustible resource.* Valuing the stock in a utilitarian framework raises the current (initial) shadow price above the classical Hotelling case, valuing the long run *à la* Chichilnisky puts the shadow price up further, and aiming at the maximum sustainable utility level raises the shadow price yet higher. This is not surprising: higher prices usually imply less consumption, and that is precisely the content of sustainability. Note, of course, that the value placed on the stock through its incorporation as an argument of the objective function is reflected in the price of the flow derived from it.

13.1.2 *Final Shadow Prices* — Can we be exact about the difference that the criterion function makes to the shadow price of the resource? We can derive analytical results for this effect in the case of the asymptotic shadow prices attained by an optimal path at a stationary configuration. Consider the shadow prices in the stationary states to which the solutions asymptote in the case in which both stock and flow are arguments of the utility function. In the utilitarian case we have

$$\lambda_{\text{util}} = u_1'(0) = u_2'/\delta \tag{13.1}$$

as the asymptotic shadow price of the resource.

If we use the Chichilnisky criterion as a maximand, then the economy does not reach the stationary state at which $u_2' = \delta u_1'$. The stationary state now satisfies (6.4), which is repeated here:

$$\frac{u_2'(\hat{s})}{u_1'(0)} = \frac{\alpha^2}{(1 - \alpha)\delta + \alpha}$$

Recall that at each point in time, the shadow price of the resource is equal to the marginal utility of consumption, from equation (4.3) of chapter 4. In this case, we have from equation (6.3) of chapter 6 that

$$-\alpha u_1'(0) + \alpha u_2'(\hat{s})/\delta + (1 - \alpha)u_2'(\hat{s}) = 0$$

or

$$\hat{\lambda} = u_2'(\hat{s})/\delta + u_2'(\hat{s})(1 - \alpha)/\alpha \tag{13.2}$$

where $\hat{\lambda}$ is the shadow price of the resource when the stationary state is reached. This differs from the utilitarian case (13.1) by the inclusion of the second term $u_2'(\hat{s})(1 - \alpha)/\alpha$. This represents the contribution of an increase in the stock to the limiting utility level, which is given positive weight by Chichilnisky's criterion. The first term of course represents the contribution of the stock to the discounted sum of utilities. The second term, although it represents contributions in the far distant future, is *not discounted*. For small values of parameter α, it can clearly be the dominant term; this corresponds to the case when the objective places more weight on the very long run than on the more immediate future.

In conclusion, the shadow price of an exhaustible resource is sensitive to the specification of the objective, increasing as more weight is placed both on the importance of the future and on the importance of the stock.

13.2   Renewable Resources

The relationships between shadow prices and the objective and discount rate are rather different in the case of renewable resources. One difference is that it does not seem possible to derive general analytical results relating the initial shadow price to the objective or to the discount rate; this is a matter of the complexity of the problem. However, it is possible to obtain results on the relationship between the objective and the terminal or stationary state shadow price.

Consider first the discounted utilitarian case with renewable resources, as formulated in chapter 4: a stationary solution satisfies (4.4), reproduced here:

$$\frac{u_2'(s_t)}{u_1'(c_t)} = \delta - r'(s_t)$$

Treating this as an implicit function and differentiating to find the relationship between c_t and δ across stationary states with different values of the discount rate, we find that if $r'(s) < 0$, then

$$\left.\frac{dc}{d\delta}\right|_{\text{stationary states}} = \frac{\partial c}{\partial \delta} + \frac{\partial c}{\partial s}\frac{\partial s}{\partial \delta} > 0$$

As the stationary states occur at a level of the stock s at which $r(s)$ is a declining function, this implies that stationary consumption decreases as the discount rate falls, which in turn implies from the equality of the shadow price of the resource and the marginal utility of consumption that the stationary state shadow price of the resource rises as the discount rate falls.

Paths that are optimal for the Chichilnisky criterion and optimize limiting utility values both have the same stationary state, namely the green golden rule. This satisfies

$$\frac{u_2'(s_t)}{u_1'(c_t)} = r'(s_t)$$

and it is immediate that here consumption is less and the resource stock greater than at any utilitarian stationary state with a positive discount rate; hence, the resource shadow price is greater. Summarizing these observations gives the following proposition:

PROPOSITION 52 Consider the problem of optimizing the time path of consumption of a renewable resource. The constraint on resource consumption is

$$\dot{s}_t = r(s_t) - c_t$$

where, as before s_t, is the resource stock at date t, and c_t the consumption of the resource at t. The objective function may be either

$$\lim_{t \to \infty} u(c_t, s_t) \text{ (green golden rule, GGR), or}$$

$$\int_0^\infty u(c_t, s_t) e^{-\delta t} \, dt + \lim_{t \to \infty} u(c_t, s_t) \text{ (Chichilnisky, Ch), or}$$

$$\int_0^\infty u(c_t, s_t) e^{-\delta t} \, dt \text{ (discounted utilitarian, DU), or}$$

$$\int_0^\infty u(c_t) e^{-\delta t} \, dt \text{(Hotelling, H)}$$

Then the shadow prices of the resource in the various stationary states, denoted λ_{ss}, satisfy the following inequalities:

$$\lambda_{ss}(\text{GGR}) = \lambda_{ss}(\text{Ch}) > \lambda_{ss}(\text{DU}) > \lambda_{ss}(\text{H})$$

Furthermore, for a discounted utilitarian solution with a constant discount rate, the stationary shadow price decreases as the discount rate increases.

Table 13.2 summarizes these conclusions.

Table 13.2

Case	Values Short Run	Values Long Run	Values Assets	Long Run Stock/Price
1. Utilitarian	Yes	No	No	Lowest
2. Stock valued	Yes	No	Yes	High
3. Chichilnisky	Yes	Yes	Yes	Higher
4. GGR	No	Yes	Yes	As in 3

Recall that in addition to the above properties, project evaluation with the Chichilnisky criterion in the case of renewable resources has a very important characteristic: it must be carried out with a time-dependent discount rate that declines to zero in the long run. In this case, the Chichilnisky optima coincide with the discounted utilitarian optima with the same timepath of the discount rate.

13.3 Sustainable Net Benefits

A new aspect of project evaluation arises in the renewable resource case. Suppose that our objective is to achieve the maximum sustainable utility, the solution to which is to move to the green golden rule. What prices, and what private behavior by agents in the economy, support such an outcome?

We noted in chapter 12 that if society is so future oriented as to wish to support the highest sustainable utility level, then we need correspondingly future-oriented behavior on the parts of agents in the economy. We need firms to seek the highest sustainable profits (i.e., the maximum value of profits that can be maintained forever) and resource owners to manage their resources so as to yield the highest sustainable revenues from the resources. In this case, we need a corresponding modification of the rules for project selection; the best of a set of projects is the one that generates the largest sustainable net benefit. In other words, the rule for a small policy change (a project) to contribute positively to welfare is that it leads to a positive indefinitely maintainable (sustainable) net benefit at the prices that support (in the above sense) the highest sustainable utility level. And the best of a set of possible projects is that which makes the largest such contribution. Note that these rules apply only in the green golden rule, and therefore in particular in a stationary state.

13.4   Investing in a Backstop Technology

Consider now the case reviewed in chapter 8. This was the case of an investment to be made today in a technology whose payoff will come only in the far future. We noted in chapter 8 that in this case the discounted utilitarian approach may recommend little or no investment in the new technology, whereas maximization of long-run utility requires the maximum possible such investment, and Chichilnisky's criterion implies an intermediate solution.

For this case, we can again establish inequalities between the shadow prices corresponding to the alternative objective functions. In particular, we can show, as for the case of depletion of an exhaustible resource, that a stronger future orientation in the objective implies a higher initial shadow price for the resource. In particular, the initial shadow price of the resource is highest with the objective of maximizing the long-run utility level, is lower than this with the Chichilnisky criterion, and is lower still with the objective of maximizing the integral of discounted utilities.

Note that in this case the Hicksian national income in this economy is zero up to the time when the new technology is available, and thereafter is given by the flow of the substitute, multiplied by the shadow price of the resource. So Hicksian national income in the long run is determined by the initial investment.

In the case of the analysis of the optimal strategy for investing in a backstop technology, this model provides an interesting insight into project evaluation according to a very long-run oriented criterion. From proposition 30 in chapter 8, the marginal return to investment in the backstop technology, within the utilitarian framework, is given by

$$\frac{\partial V_1}{\partial b}b'e^{-\delta T} + V_2'b'e^{-\delta T}$$

In the Chichilnisky framework, from the argument leading to proposition 32, the equivalent return is

$$\alpha\frac{\partial V_1}{\partial b}b'e^{-\delta T} + \alpha V_2'b'e^{-\delta T} + (1-\alpha)u'b'$$

There is a key difference between these two terms. The second contains the term $u'b'$, which is *undiscounted*: it is independent of the discount rate and of the time horizon. It represents the marginal contribution of investment in the backstop technology to the long-run or sustainable utility level, as represented by the limiting utility term in Chichilnisky's formulation. This is exactly as in (13.2). So within that framework, when we evaluate an investment project such as that modeled by our backstop model, it is appropriate to include terms representing the very long-term impact of the project that are not discounted and are therefore unaffected by changes in the discount rate or in the precise timing of any benefits. These terms measure the impact of the project on sustainable welfare. In chapter 8, we noted that current policy issues whose structures resemble that of the model in chapter 8 include managing the disposal of nuclear waste and taking measures to avoid or minimize global warming. The implication is that in both of these cases, if we accept the Chichilnisky criterion and place positive weight on sustainable utility, then it is correct to include in the benefits from such measures undiscounted terms reflecting the contributions that the measures make to welfare in the very long term. Interestingly, this provides a formal justification for an intuition emphasized by both Cline [34] and Broome [17] in their studies of global warming.

13.5 Conclusions

We identified in the earlier chapters two of the principal areas in which economic decision-making as customarily practiced, and cost–benefit analysis in particular, is deficient with respect to environmental matters. One is the valuation of the future relative to the present. The very long time horizons of many environmental preservation projects strain the credibility of discounted utilitarianism. This approach does not seem to reflect adequately the values to which societies and individuals profess. The second area of deficiency is the valuation of environmental assets, which are important sources of welfare, providing the ecological infrastructure on which human societies operate.

Several steps are required to correct these deficiencies. The first stage is the development of an appropriate theory of optimal economic development, because the theory underlying cost–benefit analysis is that of optimal economic development. The analytical relationships governing the calculation and use of shadow prices in cost–benefit analysis are drawn from this body of theory. The second stage is the embedding of principles and analytical relationships from the theory of optimal development in a set of rules for evaluating benefits and costs. The development of an appropriate intellectual framework for the theory of optimal economic development was begun in chapters 3 to 10; the embedding of principles and analytical relationships from the theory of optimal development in a set of rules for evaluating costs and benefits has been the theme of this chapter and chapter 12.

We now have a conceptual basis for beginning both of these tasks. We have a framework within which we can analyze rigorously the question "What is the optimal timepath of economic development in the presence of environmental resources, either exhaustible or renewable?" while respecting the need to place more weight on the future than is customary, and while attributing value to environmental stocks. Several clear and distinctive conclusions have emerged. These include the following:

- It may be appropriate to use a nonconstant discount rate in valuing future costs and benefits. Specifically, it may be appropriate to discount the future at a rate that declines over time. A concrete example is logarithmic discounting.
- There are conditions under which the expression for a shadow price may contain terms reflecting contributions to welfare that are completely undiscounted despite referring to benefits in the distant future.
- The precise shadow price to be applied to an environmental resource is a function of the relative weighting of present and future, and of the extent to which stocks are recognized as a source of value. This is illustrated by the results in propositions 51 and 52, which show this dependence for the cases of exhaustible and renewable resources, respectively.

Obviously, the greater part of the work in this area remains to be done. But it is clear that, in principle, what we want to do can be done. It is also clear that the emerging results are intuitively reasonable.

Appendix

This appendix focuses on an issue that is more technical than those addressed earlier and gives conditions that are sufficient for the existence of solutions to the dynamic optimization problems considered in the text. The cases of exhaustible and renewable resources are different and must be considered separately; the exhaustible case is considered first.

A.1 Existence of a Solution for the Exhaustible Resource Case

At issue here is the dynamic optimization problem of chapter 6:

$$\max \alpha \int_0^\infty u(c_t, s_t)e^{-\delta t}\, dt + (1 - \alpha) \lim_{t \to \infty} u(c_t, s_t)$$

$$\text{s.t.} \quad \dot{s}_t = -c_t, \qquad s_t \geq 0 \,\forall\, t \tag{A.1}$$

Recall that to solve this problem we defined

$$W(\beta s_0) = \max \int_0^\infty u(c_t, s_t)e^{-\delta t}\, dt$$

$$\text{s.t.} \quad \int_0^\infty c_t\, dt \leq \beta s_0, \qquad 0 \leq \beta \leq 1$$

and established certain key properties of this function in chapter 6. With this definition, the overall problem (A.1) can now be solved as follows:

$$\max_{\beta \in [0,1]} \alpha W(\beta s_0) + \begin{cases} (1 - \alpha)u_2[(1 - \beta)\, s_0] & \text{if } \beta < \beta^* \\ (1 - \alpha)u_2(s^*) & \text{otherwise} \end{cases} \tag{A.2}$$

We consider the space of possible paths of resource consumption and the resource stock as elements of the space weighted L_2, the space of functions square integrable against a finite measure, as in Chichilnisky [19]. We use the following condition in the proof:

$u(c, s)$ is concave, increasing, and differentiable. It satisfies the Caratheodory condition; namely, it is continuous with respect to c and s for almost all t and measurable with respect to t for all values of c and s.

We also use the following lemma due to Chichilnisky [19]:

LEMMA 53 (Chichilnisky) Let $W = \int_{\mathcal{R}} u(c_t, t) \, d\nu(t)$ for a finite measure $\nu(t)$, with $u(c_t, t)$ satisfying the Caratheodory condition. Then W defines a norm-continuous function from L_p to $\mathcal{R}$ for some coordinate system of L_p if and only if $|u(c_t, t)| \leq a(t) + b|c_t|^p$, where $a(t) \geq 0$, $\int_{\mathcal{R}} a(t) \, d\nu(t) < \infty$, $1 < p < \infty$, and $b > 0$.

In the case of our objective the role of the finite measure is played by $e^{-\delta t}$. An extension of Chichilnisky's lemma to functions u defined on $\mathcal{R}^2$ is straightforward. The following intermediate results are also straightforward:

LEMMA 54 If $\int_0^\infty u(c_t, s_t) e^{-\delta t} dt$ is continuous in the weighted L_2 norm, then there exists a solution to the problem $\max \int_0^\infty u(c_t, s_t) e^{-\delta t} dt$ s.t. $\int_0^\infty c_t \, dt \leq \beta s_0, 0 \leq \beta \leq 1$.

PROOF This is an immediate application of theorem 1 of Heal [58], p. 36.

It follows that if $\int_0^\infty u(c_t, s_t) e^{-\delta t} dt$ is continuous in the weighted L_2 norm, then the function $W(\beta s_0)$ is well defined. We now establish that $W(\beta s_0)$ is continuous:

LEMMA 55 The state valuation function $W(\beta s_0) = \max \int_0^\infty u(c_t, s_t) e^{-\delta t} dt$ s.t. $\int_0^\infty c_t \, dt \leq \beta s_0, 0 \leq \beta \leq 1$, is continuous as a function of the stock s_0 and parameter β.

PROOF This follows immediately from the first-order conditions for optimality in the problem defining $W(\beta s_0)$.

We now have all the preliminary results needed to prove the main result of this section, namely, the existence of a solution to the optimization problem (A.2):

PROPOSITION 56 Let $u(c, s)$ be concave, increasing, and differentiable and satisfy the Caratheodory condition. Let $|u(c_t, s_t, t)| \leq a(t) + b\|c_t, s_t\|$, where $\|x\|$ denotes the Euclidean norm of x, $a(t) \geq 0$, $\int_{\Re} a(t) e^{-\delta t} dt < \infty$ and $b > 0$. Then a solution exists to the problem (A.1) of picking an optimal usage path for the exhaustible resource with Chichilnisky's criterion.

PROOF By lemma 53, $\int_0^\infty u(c_t, s_t) e^{-\delta t} dt$ is continuous in the L_2 norm. By lemma 54, the function $W(\beta s_0)$ is well defined, and by lemma 55 it is continuous. Hence, in problem (A.2), which is equivalent to problem (A.1), the maximand is continuous with respect to β and the feasible set for β is compact. This completes the proof.

A.2 Existence of a Solution for the Renewable Resource Case

In this section we establish conditions sufficient for the existence of solutions to the various intertemporal optimization problems considered in chapter 10. These were problems involving capital accumulation and a renewable resource. We use an approach and a set of results developed initially in Chichilnisky [19] and in Chichilnisky and Kalman [33] and applied in Chichilnisky and Heal [27] and Chichilnisky and Gruenwald [25]. This is a very direct and intuitive approach: we show that the set of feasible solutions to the constraints is a compact set, and that the objective function is a continuous function, and then invoke the standard result that a continuous function on a compact set attains a maximum. The delicate step here is to find a topology in which we have compactness and continuity under reasonable assumptions about the problem. For this we use weighted L_p spaces, as introduced in Chichilnisky [19].

The optimization problem is

$$\max \int_0^\infty u(c_t, s_t) \Delta(t) dt \quad \text{s.t.}$$

$$\dot{k} = F(k, \sigma) - c \quad \text{and} \quad \dot{s} = r(s) - \sigma \tag{A.3}$$

Here $\Delta(t)$ is a discount factor whose rate of change (i.e., discount rate) goes to zero as $t \to \infty$; of course, $\int_0^\infty \Delta(t) dt < \infty$. We make the following assumptions:

1. $u(c, s)$ is concave, increasing, and differentiable. It satisfies the Caratheodory condition; namely, it is continuous with respect to c and s for almost all t and measurable with respect to t for all values of c and s.
2. $r(0) = 0$, $\exists \bar{s} > 0$ s.t. $r(s) = 0 \, \forall s \geq \bar{s}$, $\max_s r(s) \leq b_1 < \infty$, and $r(s)$ is concave for $s \in [0, \bar{s}]$.
3. For any $\sigma \exists b_2(\sigma) < \infty$ s.t. $F(k, \sigma) \leq b_2(\sigma)$.
4. $\exists b_3 < \infty$ s.t. $|\dot{s}| \leq b_3$.
5. $\exists b_4 < \infty$ s.t. $|\dot{k}| \leq b_4$.

The first two conditions are conventional. The third implies that bounded resource availability implies bounded output: it is a form of the assumption made by Dasgupta and Heal [40] that the resource is essential to production. It is a restatement of assumption (10.14). The final two assumptions imply that it is not possible for either the resource stock or the capital stock to change infinitely rapidly. These seem to be very reasonable assumptions. However, in the end we do not require them; we can prove the existence of an optimal path under these assumptions, and then note that a path that is optimal without these assumptions is still feasible and optimal with them.

PROPOSITION 57 Under the assumptions listed in this section, the utilitarian optimization problem (A.3) has a solution.

PROOF Under the assumptions, the set of feasible time paths of the resource stock s and consumption c are uniformly bounded above. (Note that s is bounded by assumption 2, and c by assumptions 3 and 5.) They are nonnegative and so bounded below. Hence, the paths of s and c are integrable against the finite measure $\Delta(t)$ and so are elements of a weighted L_1 space of functions integrable against this measure. We use the topology of this space. Denote by P the set of feasible paths s_t and c_t, $0 \leq t \leq \infty$: P is closed and norm bounded, so that by the Banach Alaoglu theorem it is weak-* compact. By Lebesgue's bounded convergence theorem, it is also compact in the norm of L_1.

The objective $U = \int_0^\infty u(c_t, s_t) \Delta(t) \, dt$ maps P to the real line $\mathfrak{R}$. To complete the proof we need to show that U is continuous in the norm of L_1. This follows immediately from the characterization of L_p continuity given in Chichilnisky [19].

LEMMA 58 (Chichilnisky) Let $W = \int_{\mathfrak{R}} u(c_t, t) \, d\nu(t)$ for a finite measure $\nu(t)$, with $u(c_t, t)$ satisfying the Caratheodory condition. Then W defines a norm-continuous function from L_p to $\mathfrak{R}$ for some coordinate system of L_p if and only if $|u(c_t, t)| \leq a(t) + b|c_t|^p$, where $a(t) \geq 0$, $\int_{\mathfrak{R}} a(t) \, d\nu(t) < \infty$, and $b > 0$.

In the case of our objective, the role of the finite measure is played by $\Delta(t)$. An extension of Chichilnisky's lemma to functions u defined on $\mathfrak{R}^2$ is straightforward. As u is defined only on $\mathfrak{R}^2_+$, concavity implies that Chichilnisky's inequality is satisfied for $p = 1$. This completes the proof of existence of an optimum.

We have now proved the existence of an optimal path for the most complex optimization problems discussed in the text; existence of an optimum for the simpler problems can be deduced from this. Our proof used assumptions 4 and 5, which bound respectively $\dot{s}$ and $\dot{k}$, the rates of change of the resource and capital stocks. These assumptions were not made in the text. However, note from the characterization results that solutions to the problems without bounds on the rates of change of stocks do have bounded rates of change of the stocks of the resource and capital. Hence, for sufficiently large bounds, the imposition of bounds on the rates of change of stocks cannot change the solutions to the optimization problems. It follows that we have also established the existence of solutions for the unbounded optimization problems.

References

[1] Ainslie, George and Nick Haslam. "Hyperbolic discounting." In George Lowenstein and Jon Elster, eds., *Choice Over Time*. New York: Russell Sage Foundation, 1992.

[2] Arrow, Kenneth J. and Frank H. Hahn. *General Competitive Analysis*. San Francisco: Holden Day, 1971, and Amsterdam: North Holland Publishing Company, 1978.

[3] Arrow, Kenneth J. and Leonid Hurwicz. "Decentralization and computation in resource allocation." In R. Pfouts, ed., *Essays in Economics and Econometrics in Honor of Harold Hotelling*. Chapel Hill: University of North Carolina Press, 1960.

[4] Asheim, Geir B. "Rawlsian intergenerational justice as a Markov-perfect equilibrium in a resource technology." *Review of Economic Studies* (1988), 50: 469–484.

[5] Asheim, Geir B. "Net national product as an indicator of sustainability." *Scandinavian Journal of Economics* (1994), 96 (2): 257–265.

[6] Asheim, Geir B. "Ethical preferences in the presence of resource constraints." *Nordic Journal of Political Economy* (1996), 23 (1): 55–67.

[7] Asheim, Geir B. "Individual and collective time consistency." *Review of Economic Studies* (1997), 64 (3): 427–443.

[8] Asheim, Geir B. "The Weitzman foundation of NNP with non-constant interest rates," June 1996. Also published as "Adjusting green NNP to measure sustainability." *Scandinavian Journal of Economics* (1997), 99 (3): 355–370.

[9] Asheim, Geir B. and Kjell A. Brekke. "Sustainability when capital management has stochastic consequences." Memorandum No. 9, 1997, Department of Economics, University of Oslo.

[10] Baskin, Yvonne. *The Works of Nature*. Washington, DC: Island Press, 1977.

[11] Beltratti, Andrea, Graciela Chichilnisky, and Geoffrey M. Heal. "Sustainable growth and the green golden rule." In I. Goldin and L. A. Winters, eds., *Approaches to Sustainable Economic Development*, pp. 147–172. Paris: Cambridge University Press for the OECD, 1993.

[12] Beltratti, Andrea, Graciela Chichilnisky, and Geoffrey M. Heal. "The environment and the long run: a comparison of different criteria." *Ricerche Economiche* (1994), 48: 319–340.

[13] Beltratti, Andrea, Graciela Chichilnisky, and Geoffrey M. Heal. "The green golden rule." *Economics Letters* (1995), 49: 175–179.

[14] Beltratti, Andrea, Graciela Chichilnisky, and Geoffrey M. Heal. "Uncertain future preferences and conservation." In G. Chichilnisky, G. M. Heal, and A. Vercelli, eds., *Sustainability: Dynamics and Uncertainty*, Chapter 3.4, pp. 255–273. Amsterdam: Kluwer Academic Publishers, Vol. 9 of *Energy Economics and Environment*, published for Fondazione ENI Enrico Mattei, 1998.

[15] Benaïm, M. and Morris W. Hirsch. "Asymptotic pseudotrajectories, chain recurrent flows and stochastic approximations." Working paper, Department of Mathematics, University of California at Berkeley, August 1994.

[16] Berry, R. Stephen, Geoffrey Heal, and Peter Salamon. "On a relation between economic and thermodynamic optima." *Resources and Energy* (1978), 1: 125–37.

[17] Broome, John. *Counting the Cost of Global Warming.* London: White Horse Press, 1992.

[18] Chichilnisky, Graciela. "Economic development and efficiency criteria in the satisfaction of basic needs." *Applied Mathematical Modeling* (September 1977), vol. 1.

[19] Chichilnisky, Graciela. "Nonlinear functional analysis and optimal economic growth." *Journal of Mathematical Analysis and Applications* (1977), 61 (2): 504–520.

[20] Chichilnisky, Graciela. "Social choice and the topology of preference spaces." *Advances in Mathematics* (1980), 37 (2): 165–176.

[21] Chichilnisky, Graciela. "Property rights and the pharmaceutical industry: a case study of Merck and InBio." Case study, Columbia Business School, New York, 1993.

[22] Chichilnisky, Graciela. "What is sustainable development?" Paper presented at Stanford Institute for Theoretical Economics, 1993. Published as "An axiomatic approach to sustainable development." *Social Choice and Welfare* (1996), 13 (2): 219–248.

[23] Chichilnisky, Graciela. "Resilience of economic and ecological systems." Working paper, Columbia University, 1996.

[24] Chichilnisky, Graciela, Vivien Gornitz, Geoffrey Heal, David Rind, and Cynthia Rosenzweig. "Building linkages between climate, impacts and economics: a new approach to integrated assessment." Working paper, NASA Goddard Institute for Space Studies, New York, 1996.

[25] Chichilnisky, Graciela and Paul Gruenwald. "Existence of an optimal growth path with endogenous technical change." *Economics Letters* (1995), 48: 433–439.

[26] Chichilnisky, Graciela and Geoffrey Heal. "Community preference and social choice." *Journal of Mathematical Economics* (1983), 12 (1): 33–62.

[27] Chichilnisky, Graciela and Geoffrey Heal. "Competitive equilibria in Sobolev spaces without bounds on short sales." *Journal of Economic Theory* (April 1983), 59 (2): 364–384.

[28] Chichilnisky, Graciela and Geoffrey Heal. "Global environmental risks." *Journal of Economic Perspectives* (Fall 1993), 7 (4): 65–86.

[29] Chichilnisky, Graciela and Geoffrey Heal. "Who should abate carbon emissions? An international perspective." *Economics Letters* (1994), 44: 443–449.

[30] Chichilnisky, Graciela and Geoffrey Heal. "Economic returns from the biosphere." *Nature* 391 (February 1998).

[31] Chichilnisky, Graciela, Geoffrey Heal, and David Starrett. "Equity and efficiency in environmental markets: global trade in CO_2 emission." Forthcoming in G. Chichilnisky and G. M. Heal, eds., *Environmental Markets*. New York: Columbia University Press, 1998.

[32] Chichilnisky, Graciela, Geoffrey Heal, and Alesandro Vercelli, eds. *Sustainability: Dynamics and Uncertainty*. Amsterdam: Kluwer Academic Publishers, for Fondazione ENI Enrico Mattei, 1997.

[33] Chichilnisky, Graciela and Peter Kalman. "Application of functional analysis to models of efficient allocation of economic resources." *Journal of Optimization Theory and Applications* (January 1980), 30 (1): 19–32.

[34] Cline, William R. *The Economics of Global Warming*. Washington, DC: Institute for International Economics, 1992.

[35] Costanza, Robert, et al. "The value of the world's ecosystem services and natural capital." *Nature* (15 May 1997), 387 (6230).

[36] Cropper, Maureen L., Sema K. Aydede, and Paul R. Portney. "Preferences for life-saving programs: how the public discounts time and age." *Journal of Risk and Uncertainty* (1994), 8: 243–265.

[37] Daily, Gretchen C., ed. *Nature's Services: Societal Dependence on Natural Ecosystems.* Washington, DC: Island Press, 1977.

[38] Daley, Herman E. *Steady State Economics: Second Edition with New Essays.* Washington, DC: Island Press, 1991.

[39] Dasgupta, Partha S. "Optimal development and the idea of net national product." In I. Goldin and L.A. Winters, eds. *Approaches to Sustainable Economic Development,* pp. 111–143. Paris: Cambridge University Press for the OECD, 1993.

[40] Dasgupta, Partha S. and Geoffrey M. Heal. "The optimal depletion of exhaustible resources." *Review of Economic Studies,* symposium (1974), pp. 3–28.

[41] Dasgupta, Partha S. and Geoffrey M. Heal. *Economic Theory and Exhaustible Resources.* Cambridge, England: Cambridge University Press, 1979.

[42] Dasgupta, Partha S., Geoffrey M. Heal, and Mukul K. Majumdar. "Resource depletion and research and development." In M. D. Intriligator, ed., *Frontiers of Quantitative Economics,* vol. IIIB, pp. 483–505. Amsterdam: North Holland Publishing, 1976.

[43] Dasgupta, Partha S., Geoffrey M. Heal, and Anand K. Pand. "Funding research and development." *Applied Mathematical Modeling* (April 1980), 4: 87–94.

[44] Dasgupta, Partha S., Bengt Kriström, and Karl-Göran Mäler. "Current issues in resource accounting." In P.-O. Johansson, B. Kriström, and K.-G. Mäler, eds., *Current Issues in Environmental Economics.* Manchester, England: Manchester University Press.

[45] Dasgupta, Partha S., Steven Marglin, and Amartya Sen. *Guidelines for Project Evaluation.* New York: United Nations, 1972.

[46] Dixit, Avinash K., Peter Hammond, and Michael Hoel. "On Hartwick's rule for regular maximin paths of capital accumulation and resource depletion." *Review of Economic Studies* (April 1980), 47 (3), 148: 551–556.

[47] Dutta, Prajit. "What do discounted optima converge to? A theory of discount rate asymptotics in economic models." *Journal of Economic Theory* (October 1991), 55 (1): 64–94.

[48] Fisher, Irving. *The Nature of Capital and Income.* New York: Macmillan, 1906.

[49] Frerejohn, John and Talbot Page. "On the foundations of intertemporal choice." *Journal of Agricultural Economics* (May 1978), pp. 15–21.

[50] Graaff, J. de V. *Theoretical Welfare Economics.* Cambridge, England: Cambridge University Press, 1963.

[51] Gurr, Sarah Jane and Josephine Peach. "The hidden power of plants." *Journal of the Royal Horticultural Society* (May 1996), 121 (5): 262–264.

[52] Hammond, Peter J. and James A. Mirrlees, "Agreeable plans." In J. A. Mirrlees and N. Stern, eds., *The Theory of Economic Growth.* New York: Macmillan, 1973.

[53] Harrod, Roy. *Towards a Dynamic Economics.* London: Macmillan, 1948.

[54] Hartwick, John M. "Intergenerational equity and investing the rents from exhaustible resources." *American Economic Review* (1977), 66: 9072–9074.

[55] Hartwick, John M. "National wealth and net national product." *Scandinavian Journal of Economics* (1994) 96 (2): 253–256.

[56] Harvey, Charles. "The reasonableness of non-constant discounting." *Journal of Public Economics* (1994), 53: 31–51.

[57] Heal, Geoffrey M. *The Theory of Economic Planning,* Amsterdam: North Holland Publishing, 1973.

[58] Heal, Geoffrey M. "Depletion and discounting: a classical issue in resource economics." In R. McElvey, ed., *Environmental and*

Natural Resource Mathematics (1985), vol. 32, pp. 33–43, Proceedings of Symposia in Applied Mathematics, American Mathematical Society, Providence, RI.

[59] Heal, Geoffrey M. *The Economics of Exhaustible Resources.* International Library of Critical Writings in Economics, Edward Elgar, 1993.

[60] Heal, Geoffrey M. "The optimal use of exhaustible resources." In Alan V. Kneese and James L. Sweeney, eds., *Handbook of Natural Resource and Energy Economics,* ch. 18, pp. 855–880. Amsterdam: North Holland, 1993.

[61] Heal, Geoffrey M. "Valuing the very long run." Working paper, Columbia Business School, New York, December 1993.

[62] Herrera, Amilcar O., Hugo D. Scolnik, Graciela Chichilnisky, et al. *Catastrophe or New Society.* Ottawa: International Development Research Centre, 1976.

[63] Hicks, John R. *Value and Capital,* 2nd ed. New York: Oxford University Press, 1939.

[64] Holdren, John P., Gretchen C. Daily, and Paul R. Ehrlich. "The meaning of sustainability: biogeophysical aspects." In Mohan Munasinghe and Walter Shearer, eds., *Defining and Measuring Sustainability: The Biogeophysical Foundations.* Distributed for the United Nations University by the World Bank, Washington, DC, 1995.

[65] Hollings, C. S. "Resilience and stability of ecological systems." *Annual Review of Ecological Systems,* 4: 1–24.

[66] Hotelling, Harold. "The economics of exhaustible resources." *Journal of Political Economy* (1931), 39: 137–175.

[67] Kamien, Morton and Nancy Schwartz. *Dynamic Optimization.* Amsterdam: North Holland, 1991.

[68] Karp, Larry and David Newberry. "Intertemporal consistency issues in depletable resources." In Alan Kneese and James Sweeney,

eds., *Handbook of Natural Resource and Energy Economics,* vol. 3, ch. 19, pp. 881–931. Amsterdam: North Holland, 1993.

[69] Kneese, Alan V. and William D. Schultze. "Ethics and environmental economics." In Alan Kneese and James Sweeney, eds., *Handbook of Natural Resource and Energy Economics,* vol. 1. Amsterdam: North Holland, 1985.

[70] Koopmans, Tjalling. "Stationary ordinal utility and impatience." *Econometrica* (1960), 28: 287–309.

[71] Krautkraemer, Jeffrey A. "Optimal growth, resource amenities and the preservation of natural environments." *Review of Economic Studies* (1985), 52: 153–170.

[72] Krautkraemer, Jeffrey A. "Optimal depletion with resource amenities and a backstop technology." *Resources and Energy* (1986), 8: 133–149.

[73] Lauwers, Luc. "Infinite Chichilnisky rules." Discussion paper, Katolik Universitaet Leuven, Belgium, 1992.

[74] Lauwers, Luc and Luc van Liederkirke. "Monotonic Chichilnisky rules with infinite populations." Discussion paper, Katolic Universitaet Leuven, Belgium, 1992.

[75] Le Kama, Alain Ayong. "Sustainable growth, renewable resources and pollution." Working paper, MAD, CNRS, Université de Paris I, 1996.

[76] Lindahl, Erik. "The concept of income." In G. Bagge, ed., *Economic Essays in Honor of Gustav Cassell.* Allen & Unwin, 1933.

[77] Little, Ian M. D. and James A. Mirrlees. *Project Appraisal and Planning for Developing Countries.* New York: Basic Books, 1974.

[78] Lowenstein, George and Jon Elster, eds., *Choice Over Time.* New York: Russell Sage Foundation, 1992.

[79] Lowenstein, George and Drazen Prelec. "Anomalies in intertemporal choice: evidence and an interpretation." In George Lowenstein

and Jon Elster, eds., *Choice Over Time*. New York: Russell Sage Foundation, 1992.

[80] Lowenstein, George and Richard Thaler. "Intertemporal choice." *Journal of Economic Perspectives* (1989), 3: 181–193.

[81] Malinvaud, Edmond. "Capital accumulation and the efficient allocation of resources." *Econometrica* (1953), 21 (2): 233–268.

[82] Meade, James E. "The effect of savings on consumption in a state of steady growth." *Review of Economic Studies* (June 1962), 29.

[83] Murphy, Tony. "What value nature: a legal viewpoint." *Journal of Environmental Education* (Summer 1996), 27 (4): 5–8.

[84] Newman, Arnold. *Tropical Rainforests: A World Survey of Our Most Valuable and Endangered Habitat with a Blueprint for Its Survival*. New York: Facts on File, 1990.

[85] Ng, Yew-Kwang. "Towards welfare biology: evolutionary economics of animal consciousness and suffering." Working paper, Monash University, Australia 3168, forthcoming in *Biology and Philosophy*.

[86] Nordhaus, William D. "Is growth sustainable? Reflections on the concept of sustainable economic growth." Paper presented to the International Economic Association Conference, Varenna, Italy, October 1992.

[87] Nordhaus, William D. "How should we measure sustainable income?" Discussion paper, Department of Economics, Yale University, New Haven, CT, 1995.

[88] Nordhaus, William D. and James Tobin. "Is growth obsolete?" In *Economic Growth*. New York: National Bureau of Economic Research, 1972; also in *Income and Wealth*, vol. 38. New York: National Bureau of Economic Research, 1973.

[89] Pearce, David W., Anil Markandya, and Edward Barbier. *Sustainable Development: Economy and Environment in the Third World*. London: Earthscan Publications, 1990.

[90] Phelps, Edmund S. "The golden rule of accumulation: a fable for growthmen." *American Economic Review* (1961), 638–643.

[91] Ramsey, Frank. "A mathematical theory of saving." *Economic Journal* (1928), 38: 543–559.

[92] Rawls, John. *A Theory of Justice.* Oxford, England: Clarendon, 1972.

[93] Repetto, Robert, Wilfrido Cruz, Raúl Solórzano, Ronnie de Camino, Richard Woodward, Joseph Tosi, Vicente Watson, Alexis Vásquez, Carlos Villalobos, and Jorge Jiménez. "Accounts overdue: natural resource depreciation in Costa Rica." Washington, DC: World Resource Institute, and San José, Costa Rica: Tropical Science Center, 1991.

[94] Repetto, Robert, William Magrath, Michael Wells, Christine Beer, and Fabrizio Rossinni. "Wasting assets: natural resources in the national income accounts." Washington, DC: World Resources Institute, 1989; also ch. 25 in Anil Markandya and Julie Richardson, eds., *Environmental Economics: A Reader.* New York: St. Martin's Press, 1992.

[95] Robinson, Joan. "A neoclassical theorem." *Review of Economic Studies* (June 1962), 29.

[96] Rolston, H. *Environmental Ethics: Duties to and Values in the Natural World.* Philadelphia: Temple University Press, 1988.

[97] Roughgarden, Joan and Fraser Smith. "Why fisheries collapse and what to do about it." Forthcoming in the *Proceedings of the National Academy of Sciences.*

[98] Ryder, Harl Jr. and Geoffrey Heal. "Optimal growth with intertemporally dependent preferences." *Review of Economic Studies* (1973), 40 (121): 1–32.

[99] Seirstad, Atle and Knut Sydsæter. *Optimal Control Theory with Economic Applications.* Amsterdam: North Holland, 1987.

[100] Serageldin, Ismael. *Sustainability and the Wealth of Nations: First Steps on an Ongoing Journey.* Prepared for the Third Annual World Bank Conference on Environmentally Sustainable Development, World Bank, Washington, DC, September 1995.

[101] Sidgwick, H. *The Methods of Ethics.* London: Macmillan, 1890.

[102] Smith, Adam. *An Inquiry into the Nature and Causes of the Wealth of Nations.* Chicago: University of Chicago Press, 1977.

[103] Solow, Robert M. "A contribution to the theory of economic growth." *Quarterly Journal of Economics* (1956), 70 (1): 65–94.

[104] Solow, Robert M. "Intergenerational equity and exhaustible resources." *Review of Economic Studies,* **Symposium on the Economics of Exhaustible Resources** (1974), pp. 29–45.

[105] Solow Robert M. *An Almost Practical Step Towards Sustainability.* Invited lecture on the fortieth anniversary of Resources for the Future, Resources and Conservation Center, Washington, DC, 1992.

[106] Thaler, Richard. "Some empirical evidence on dynamic inconsistency." *Economics Letters* (1981), 8: 201–207.

[107] von Weizäcker, Carl Christian. "Lemmas for a theory of approximately optimal growth." *Review of Economic Studies* (1967), pp. 143–151.

[108] Wagner, Jeffrey. "Radioactive waste treatment and the natural resource economics paradigm." Working paper, University of Maryland, University College, Heidelberg, Germany.

[109] Weitzman, Martin L. "On the welfare significance of net national product in a dynamic economy." *Quarterly Journal of Economics* (1976), 90: 156–162.

[110] World Commission on Environment and Development. *Our Common Future* (the Brundtland Report). New York: Oxford University Press, 1987.

Index

Agreeable plans, 114
Ainslie, George, 62
Asheim, Geir, 4, 7, 59, 156, 162
Automonous, asymptotically, 100, 101

Banach Alaoglu Theorem, 210
Bariloche, 6, 10, 21
Baskin, Yvonne, 3, 14
Beltratti, Andrea, 10, 19
Benaim, Michel, 100, 101
Bentham, Jeremy, 12
Berry, Stephen, 148
Biodiversity, 1, 2, 5, 15, 17, 18, 19, 21, 36, 38
Brekke, Kjell, 4
Broome, John, 12, 206
Brundtland, Gro, 6, 10, 21

Catskills, 16, 17
Chichilnisky criterion, 69ff, 82ff, 94ff, 107ff, 123ff, 137ff, 141, 150ff, 170, 173, 185ff, 196ff, 203ff
Chichilnisky, Graciela, 4, 5, 8, 9, 10, 11, 14, 17, 19, 22, 23, 24, 53, 59, 69ff, 82ff, 94, 116, 123ff, 137ff, 157, 173, 195, 208ff
Cline, William, 12, 206
Confucius, 1
Consistency, asymptotic, 107ff; intertemporal, 104ff
Consumption discount rate, 76ff

Costa Rica, national income of, 195
Cropper, Maureen, 61

Daily, Gretchen, 1, 3, 14, 16, 18
Daley, Herman, 8, 11, 21
Dasgupta, Partha, 4, 7, 11, 30, 31, 47, 58, 60, 64, 65, 117, 119, 128, 129, 148, 156, 210
Discounting, logarithmic, 62ff
Discount rate: 12ff;
    applied to consumption, 76ff
    variable, 30ff, 61ff, 98ff
    zero, 63ff, 65ff, 92ff, 110, 111ff, 189ff, 204
Dixit, Avinash, 8, 10

Ecological stability, 53
Ecosystem, 19, 47, 196
Ecosystem services, 3, 24
Ehrlich, Paul, 1, 16
Elster, Jon, 61, 109

Fisher, Irving, 4, 8, 10, 23, 25, 35, 156

Global commons, 4
Global warming, 1, 2, 21, 22, 116, 196
Golden rule, 4, 10, 11, 43
Graaff, Jan de Villiers, 158
Green golden rule, 10, 21, 24, 25, 41, 43, 49, 52, 54, 59, 88, 101, 102, 111, 112, 113, 114, 122, 135, 139, 146, 152, 196ff, 204ff

Gruenwald, Paul, 209
Gurr, Sarah Jane, 15

Hammond, Peter, 8, 10, 114
Harrod, Roy, 12, 23, 58, 79
Harsanyi, John, 60
Hartwick, John, 8, 10, 156
Harvey, Charles, 63, 109
Haslam, Nick, 62
Heal, Geoffrey, 4, 5, 10,11, 12,
    17, 19, 28, 30, 31, 47, 58, 60,
    64, 65, 114, 117, 119, 128,
    129, 132, 148, 157, 193, 208,
    210
Hicks, John, 4, 7, 8, 10, 23, 25,
    35, 156ff, 180ff
Hirsch, Morris, 100, 101
Hoel, Michael, 8
Holdren, John, 1
Hollings, C., 11
Hotelling, Harold, 27, 29, 31, 42,
    43, 44, 79, 81ff
Hotelling model, 27ff, 81ff, 90,
    116, 166, 181ff, 197ff, 203ff
Hotelling rule, 29, 116, 119

Income, 4, 7, 8, 23, 25, 35;
    Hicksian income, 156ff, 180ff,
        196, 205
    national income, 25, 155ff,
        175ff;
Independence axiom, 60, 74
Indonesia, national income of,
    195
International Rice Research
    Institute, 15

Kalman, Peter, 209
Karp, Larry, 107
Keynesian, 155
Kneese, Alan, 17

Koopmans, Tjalling, 60, 61, 74,
    75
Krautkraemer, Jeffrey, 19, 36,
    129
Kristrom, Bengt, 7, 156

Lebesgue-bounded convergence
    theorem, 210
Le Kama, Alain, 64
Lindahl, Erik, 4, 7, 8, 10, 23, 25,
    35, 156
Little, Ian, 4
Logarithmic discounting, 30,
    62ff, 106, 107
Long-run average payoff, 68
Lowenstein, George, 13, 61, 62,
    109

Maler, Karl-Goran, 7, 156
Marglin, Steven, 4
Markandya, Anil, 8
Matson, Pamela, 16
Meade, James, 4, 9
Mirrlees, James, 4, 114
Murphy, Tony, 17
Myers, Norman, 16

Natural capital, 8
Naylor, Rosamund, 16
Newberry, David, 107
New York, 16, 17
Ng, Yew Kwang, 17
Nordhaus, William, 7, 158

Overtaking criterion, 65ff, 73, 92,
    93, 110, 116, 125

Pareto condition, 61, 71
Peach, Josephine, 15
Pearce, David, 8, 11, 21

Phelps, Edmund, 4, 9, 43
Prelec, Drazen, 61, 62
Public goods, 4, 5

Ramsey, Frank, 12, 23, 58, 63,
    64, 79
Rawls, John, 8, 22, 60
Rawlsian, 7, 8, 9, 11, 23, 24, 44,
    55, 58, 59, 76, 123, 153, 197
Repetto, Robert, 195
Robinson, Joan, 4, 9
Rolston, H., 17
Roughgarden, Joan, 53
Ryder, Harl, 65, 99, 103

Salamon, Peter, 148
Schultze, William, 17
Seirstad, Atle, 28
Sen, Amartya, 4
Sensitivity: to the long run, 69ff;
    to the present, 70ff
Sidgwick, Henry, 23, 58
Smith, Adam, 2
Smith, Frazer, 53
Solow, Robert, 4, 8, 9, 10, 11, 21,
    129
Starrett, David, 5
Stationarity axiom, 60, 74

Sustainability, 1, 2, 3, 4, 5, 6, 7,
    9, 13, 18, 20, 21, 31, 36, 45,
    55, 56, 200
Sustainable revenues, 189ff
Sydsaeter, Knut, 28

Technical change, 4
Thaler, Richard, 13, 61
Thermodynamics, second law of,
    148
Tobin, James, 158
Turnpike, 111, 112

Uncertainty, 4
Utilitarianism, 12ff, 60, 91, 118ff,
    129ff, 142ff

Vercelli, Alessandro, 4
Vitousek, Peter, 16
von Weizacker, Karl Christian, 23

Wagner, Jeffrey, 115
Walrasian stability, 9
Wealth, national, 175ff
Weber-Fechner law, 62, 63, 104
Weitzman, Martin, 23, 156
Wilen, James, 50

9 780231 113076